Chemistry Research and Applications

Chemistry Research and Applications

The Chemistry of Gallic Acid and Its Role in Health and Disease
Jeff C. Murdoch (Editor)
2023. ISBN: 979-8-88697-672-4 (Softcover)
2023. ISBN: 979-8-88697-729-5 (eBook)

A Review of Hydrazine and Its Applications
Thomas W. Gerdes (Editor)
2023. ISBN: 979-8-88697-671-7 (Hardcover)
2023. ISBN: 979-8-88697-678-6 (eBook)

Pyrene: Chemistry, Properties and Uses
Charles R. Howe (Editor)
2023. ISBN: 979-8-88697-670-0 (Softcover)
2023. ISBN: 979-8-88697-677-9 (eBook)

Pyrimidines and their Importance
Roger G. Ward (Editor)
2023. ISBN: 979-8-88697-656-4 (Softcover)
2023. ISBN: 979-8-88697-663-2 (eBook)

What to Know about Lanthanum
Catherine C. Bradley (Editor)
2023. ISBN: 979-8-88697-615-1 (Softcover)
2023. ISBN: 979-8-88697-623-6 (eBook)

The Future of Biorefineries
Waldemar Nyström (Editor)
2023. ISBN: 979-8-88697-524-6 (Hardcover)
2023. ISBN: 979-8-88697-528-4 (eBook)

More information about this series can be found at
https://novapublishers.com/product-category/series/chemistry-research-and-applications/

Stephen A. Reyes
Editor

Imidazolium

Properties, Applications and Biological Significance

NOTICE TO THE READER

Library of Congress Cataloging-in-Publication Data

ISBN: 979-8-89113-425-6

Published by Nova Science Publishers, Inc. † New York

Contents

Preface

This book is a compilation of five chapters. The goal of Chapter One is to organize the literature studies on thermal decomposition of Dicationic Ionic Liquids (DILs), demonstrating how decomposition is generally measured and which parameters are utilized to characterize decomposition properties. In Chapter Two, the authors are describing the synthesis of chiral ionic liquids and their application in asymmetric organic transformations. Chapter Three provides a review of G4 organization, including the structures and mechanisms for G4 detection, highlights the advantages of G4 ligand structures, and provides insights into future prospects in the field. Chapter Four is a DFT study of the different properties and applications of Imidazolium Ionic Liquids. The final chapter focuses on the recent reports on dicationic liquid crystals based on imidazolium motifs, both flexibly or rigidly linked bisimidazolium salts and will also discuss new perspectives in the design and targeted applications of the bisimidazolium based ILCs.

Chapter 1

Thermal Stability of Dicationic Imidazolium-Based Ionic Liquids

Alisson V. Paz[1]
Jean C. B. Vieira[1]
Clarissa P. Frizzo[1,*]**, PhD**
and Caroline R. Bender[1,†]**, PhD**
[1]Department of Chemistry, Federal University of Santa Maria (UFSM), Rio Grande do Sul, Brazil

Abstract

Dicationic Imidazolium-based Ionic Liquids are important in a large variety of applications. Due to their inimitable properties, they have been widely used as solvents and lubricants. The goal of this current chapter is organizing the literature studies regarding thermal decomposition of Dicationic Ionic Liquids (DILs), showing as decomposition is generally measured and which parameters are used to characterize decomposition properties. Furthermore, decomposition kinetics and mechanisms were addressed to present what is known about the subject at this time. The chapter can be helpful to the researchers to choose the more appropriate DILs to be applied in several studies considering which structural modifications in spacer length, side chains or presence of functional groups in DILs can be more advantageous. Additionally, the lack of standardizations in existing studies that limit the tendencies establishment and comparisons, and the scarce use of kinetic analysis to determine the long-term thermal stability and decomposition mechanisms of DILs were highlighted.

* Corresponding Author's Email: clarissa.frizzo@gmail.com.
† Corresponding Author's Email: carolinerbender@gmail.com.

In: Imidazolium
Editor: Stephen A. Reyes
ISBN: 979-8-89113-425-6

Keywords: dicationic Ionic liquids, thermal stability, decomposition

Abbreviation and Symbols

^{1}H-NMR	Hydrogen Nuclear Magnetic Resonance
A	Pre-exponential factor
CTf_3	Tris(trisfluoromethanesulfonyl)methide
DFT	Density functional theory
DILs	Dicationic Ionic Liquids
DSC	Differential Scanning Calorimetry
E_a	Activation energy
ESI-MS	Electrospray Ionization Mass Spectrometry
$f(\alpha)$	Theoretical kinetic model
fNTf	fluorosulfonyl)(trifluoromethanesulfonyl)imide
FTIR	Fourier Transformation Infrared Spectrometry
$g(\alpha)$	Integrated form of $f(\alpha)$
ICTAC	International Confederation for Thermal Analysis and Calorimetry
k(T)	Rate constant
KAS	Kissinger–Akahira–Sunose isoconversional method
m/z	Mass-to-charge ratio
m_f	Final mass
m_i	Initial mass
MS	Mass Spectrometry
m_T	Mass at temperature T
m_t	Mass at time t
Nf_2	Bis(fluorosulfonyl)imide
NfO	Nonafluoro-1-butanesulfonate
NPf_2	Bis(pentafluoromethanesulfonyl)imide
NTf_2	Bis(trifluoromethylsulfonyl)imide
OFW	Osawa–Fynn–Wall isoconversional method
PEG	Polyethylene glycol
q	Heating rate
R	Universal gas constant
STA	Simultaneous Thermal Analysis
t	Time
T	Temperature
T_d	Decomposition temperature

$T_{d\,max}$	Maximum decomposition of derivative graph peak
$T_{d\,n\%}$	Decomposition temperature at a given mass loss (n = 1, 5, 10, 15%)
$T_{d\,onset}$	Intersection of the baseline weight, and the tangent of the weight dependence on the temperature curve as decomposition occurs.
TfO	Trifluoromethanesulfonate (triflate)
TGA	Thermogravimetric Analysis
α	Conversion
$\Delta G^{\ddagger}$	Activation Gibbs free energy
$\Delta H^{\ddagger}$	Activation enthalpy
$\Delta S^{\ddagger}$	Activation entropy

Introduction

Ionic liquids (ILs) are organic salts with low melting points. ILs are termed "designer" owing to the great flexibility in tailoring the cationic and anionic structures and combinations according to the required properties (Frizzo et al. 2015). One of the most extensively studied ILs is the imidazolium species. Imidazolium compounds are known to exhibit high thermal stability, which is desirable for industrial applications (Plechkova and Seddon 2008). One of the most important properties of ILs is their decomposition temperature. However, this temperature is known to overestimate their thermal stability (Maton, Vos, and Stevens 2013). One way to overcome this problem is to analyze the kinetic parameters (e.g., the activation energy and pre-exponential factor) of the thermal decomposition. The two principal procedures used to determine the kinetic parameters are the isothermal and isoconversional methodologies. Kinetic analysis can also help to obtain an understanding of the decomposition mechanism (Xu and Cheng 2021). For monocationic ILs, more complete information about thermal decomposition and kinetic decomposition exists (Xu and Cheng 2021). However, there are only a few reports on the thermal degradation products of dicationic imidazolium-based ILs, particularly those with organic anions. Searching Web of Science using the term "dicationic ionic liquids" resulted in 699 references, of which 311 contained the term "imidazolium," and only 25 also included "thermal decomposition." In this context, the aim of this chapter is to systematize the studies on the thermal decomposition of dicationic imidazolium-based ILs to find some trends and present what we know about them at this point. This knowledge is of pivotal importance, considering that the use of ILs in thermally dependent processes requires knowledge of their thermal properties and stability. The thermal decomposition, methodologies used to determine the kinetic parameters (isothermal and isoconversional), and mechanism of decomposition of dicationic ILs will be presented and discussed in comparison with those of other ILs. Thus, this chapter provides a comprehensive survey of the thermal stability of dicationic imidazolium-based ILs and the structural effects of cations and anions on this property, which will simultaneously contribute to the systematization of knowledge as well as assist researchers.

Thermal Profile of Dicationic ILs

According to the International Confederation for Thermal Analysis (ICTAC), thermal analysis (TA) is the study of the relationship between a sample property and its temperature when the sample is heated or cooled in a controlled manner (Lever et al. 2014). Thermogravimetric Analysis (TGA) is a technique in which the properties of a sample are correlated with its mass.

The most popular technique used to study the thermal degradation of ionic liquids (ILs) is TGA, which involves measuring the weight loss induced by increasing the temperature in a controlled atmosphere (Wooster et al. 2006). The onset temperature (T_{onset}) is usually used to report the thermal stability or profile of ILs. This parameter represents the intersection of the baseline weight with the tangent of the weight dependence on the temperature curve during decomposition (Siedlecka et al. 2011). The assumption of this value as the decomposition temperature of an IL is useful in comparative studies, where the objective is to elucidate the structural tendencies. The structural tendencies and how they affect an ILs thermal stability are the main subjects of this section. The objective is to trace the findings reported in the existing literature to understand how structural modifications can directly affect the thermal stability of ILs and to highlight the many divergences in the data by compiling the most representative works.

The discussions in this section are based on comparing the structures and degradation temperatures compiled in Table 1. As can be seen, the authors have reported the thermal profiles in different ways: $T_{d\ max}$ represents the temperature of maximum mass loss, $T_{d\ n\%}$ is the temperature corresponding to a specific percentage of mass loss (n = 1, 5, 10, 15%), $T_{d\ final}$ is the temperature at which the decomposition appears to cease in the TGA thermogram, and $T_{d\ onset}$ has already been discussed. In cases where the authors did not mention the T_d, we determined the values from TGA curves and included them in Table 1 to facilitate comparison of the structures and add more depth to the discussion on structural diversity. At the bottom of the table, the heating rate of the experiments (°C min^{-1}) is also specified for each IL.

Table 1. Representation of ILs in this Chapter and decomposition temperature of ILs

Entry	Ionic Liquid	T_d (°C)	Reference
1	N N $^{\oplus}$ $Br^{\ominus}$	300[a] ($T_{d\,max}$)	(Cao and Mu 2014)
2	$Br^{\ominus}$ N N$^{\oplus}$ $(\;)_2$ $^{\oplus}$N N $Br^{\ominus}$	323[a] ($T_{d\,max}$)	(Zhang et al. 2019)
3	$Br^{\ominus}$ N N$^{\oplus}$ $(\;)_3$ $^{\oplus}$N N $Br^{\ominus}$	321[a] ($T_{d\,max}$)	(Zhang et al. 2019)
4	$Br^{\ominus}$ N N$^{\oplus}$ $(\;)_4$ $^{\oplus}$N N $Br^{\ominus}$	318[a] ($T_{d\,max}$)	(Zhang et al. 2019)
5	$Br^{\ominus}$ N N$^{\oplus}$ $(\;)_5$ $^{\oplus}$N N $Br^{\ominus}$	310[a] ($T_{d\,max}$)	(Zhang et al. 2019)
6	$Br^{\ominus}$ N N$^{\oplus}$ $(\;)_7$ $^{\oplus}$N N $Br^{\ominus}$	343[b] ($T_{d\,max}$)	(Claros et al. 2010)
7	$Br^{\ominus}$ N N$^{\oplus}$ $(\;)_8$ $^{\oplus}$N N $Br^{\ominus}$	338[b] ($T_{d\,max}$)	(Claros et al. 2010)
8	$Br^{\ominus}$ N N$^{\oplus}$ $(\;)_{10}$ $^{\oplus}$N N $Br^{\ominus}$	317[a] ($T_{d\,max}$)	(Bender et al. 2019)
9	$Br^{\ominus}$ N N$^{\oplus}$ $(\;)_{12}$ $^{\oplus}$N N $Br^{\ominus}$	309[a] ($T_{d\,max}$)	(Bender et al. 2019)
10	$Br^{\ominus}$ N N$^{\oplus}$ $(\;)_{14}$ $^{\oplus}$N N $Br^{\ominus}$	315[a] ($T_{d\,max}$)	(Bender et al. 2019)

Table 1. (Continued)

Entry	Ionic Liquid	T_d (°C)	Reference
11	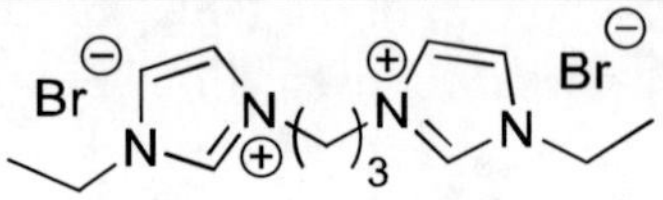	316[a] ($T_{d\,max}$)	(Zhang et al. 2019)
12	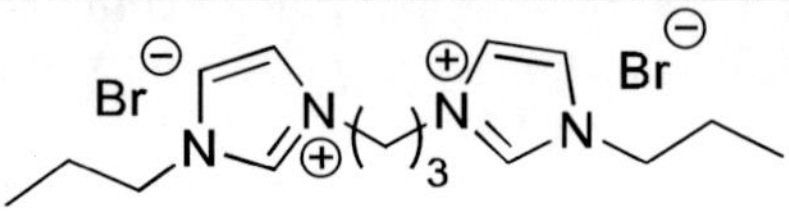	304[a] ($T_{d\,max}$)	(Zhang et al. 2019)
13	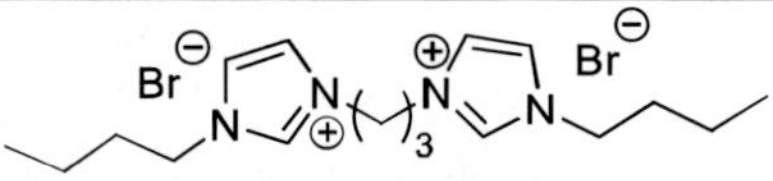	302[a] ($T_{d\,max}$)	(Zhang et al. 2019)
14	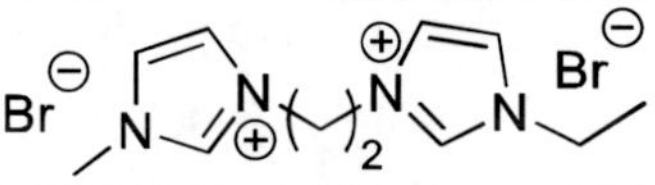	303[a] ($T_{d\,max}$)	(Zhang et al. 2019)
15	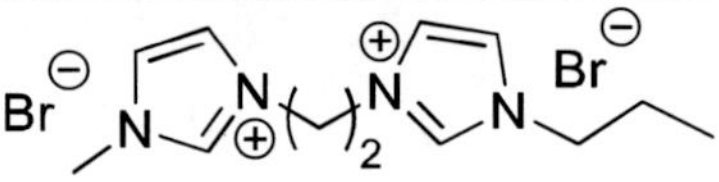	293[a] ($T_{d\,max}$)	(Zhang et al. 2019)
16	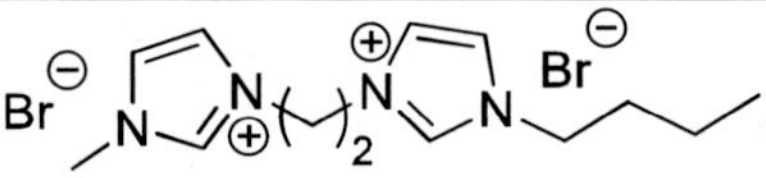	291[a] ($T_{d\,max}$)	(Zhang et al. 2019)
17	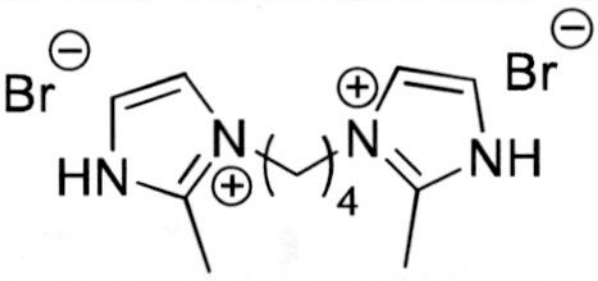	317[a] ($T_{d\,max}$)	(J. Liu et al. 2020)
18	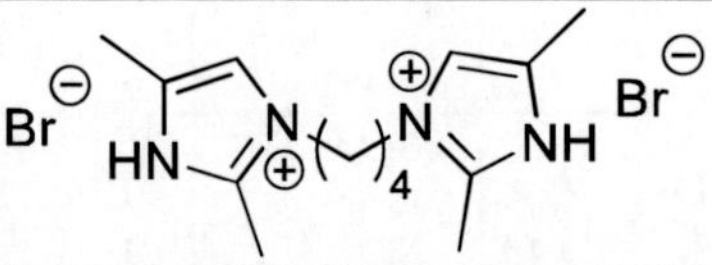	329[a] ($T_{d\,max}$)	(J. Liu et al. 2020)
19	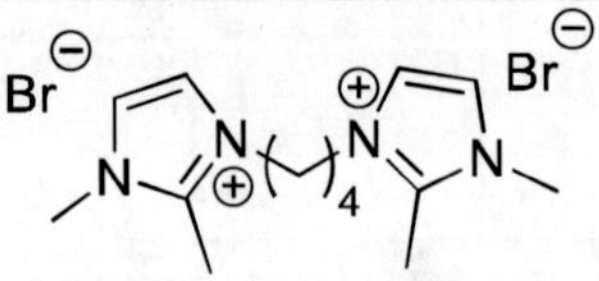	327[a] ($T_{d\,max}$)	(J. Liu et al. 2020)
20	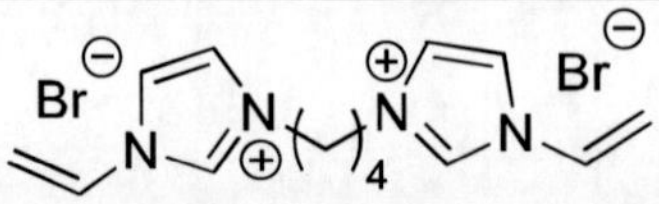	304[a] ($T_{d\,10\%}$)	(Zhang et al. 2018)
21	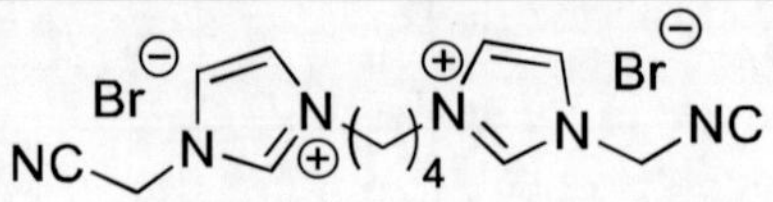	320[a] ($T_{d\,10\%}$)	(Zhang et al. 2018)

Entry	Ionic Liquid	T_d (°C)	Reference
22		309[a] ($T_{d\,10\%}$)	(Zhang et al. 2018)
23		298[a] ($T_{d\,10\%}$)	(Zhang et al. 2018)
24		475[a] ($T_{d\,15\%}$)	(Talebi, Patil, and Armstrong 2018)
25		442[a] ($T_{d\,15\%}$)	(Talebi, Patil, and Armstrong 2018)
26		478[a] ($T_{d\,15\%}$)	(Talebi, Patil, and Armstrong 2018)
27		472[a] ($T_{d\,15\%}$)	(Talebi, Patil, and Armstrong 2018)
28		475[a] ($T_{d\,15\%}$)	(Talebi, Patil, and Armstrong 2018)
29		468[a] ($T_{d\,15\%}$)	(Talebi, Patil, and Armstrong 2018)

Table 1. (Continued)

Entry	Ionic Liquid	T_d (°C)	Reference
30	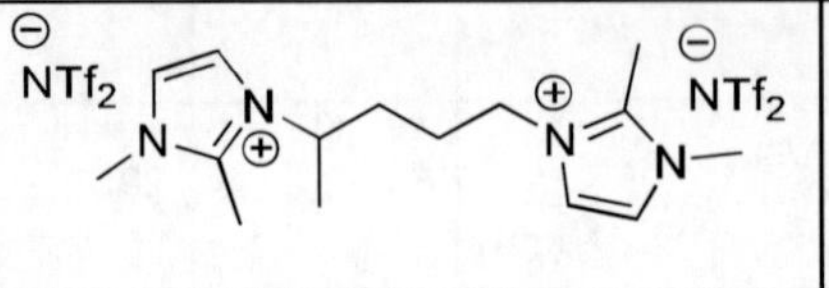 NTf_2^- NTf_2^-	425[a] ($T_{d\ 15\%}$)	(Talebi, Patil, and Armstrong 2018)
31	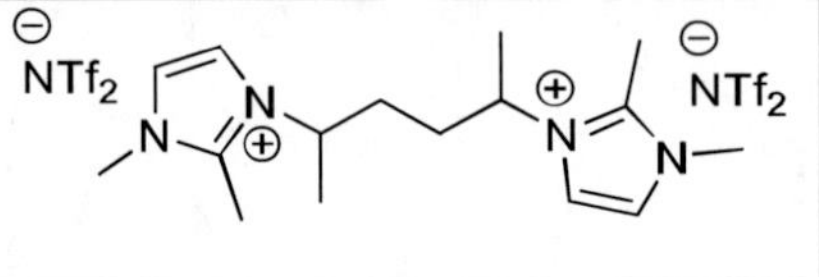 NTf_2^- NTf_2^-	407[a] ($T_{d\ 15\%}$)	(Talebi, Patil, and Armstrong 2018)
32	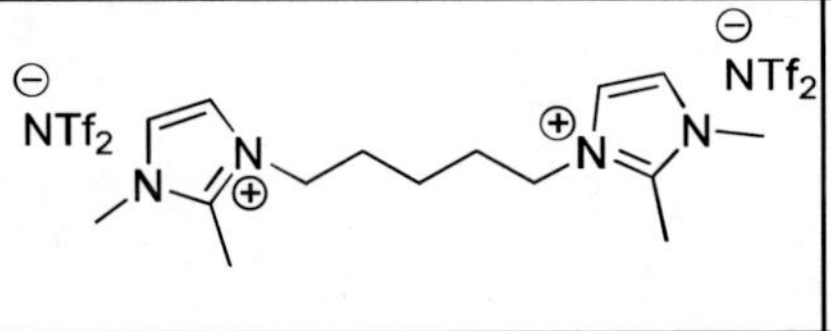 NTf_2^- NTf_2^-	477[a] ($T_{d\ 15\%}$)	(Talebi, Patil, and Armstrong 2018)
33	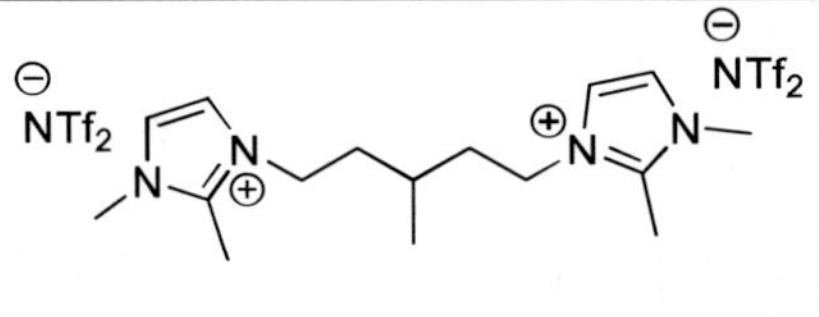 NTf_2^- NTf_2^-	464[a] ($T_{d\ 15\%}$)	(Talebi, Patil, and Armstrong 2018)
34	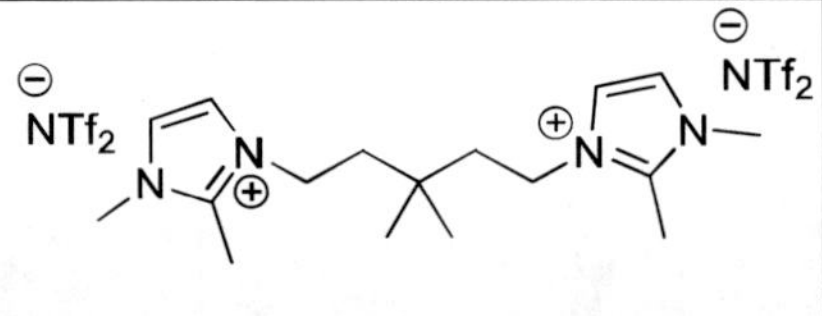 NTf_2^- NTf_2^-	446[a] ($T_{d\ 15\%}$)	(Talebi, Patil, and Armstrong 2018)
35	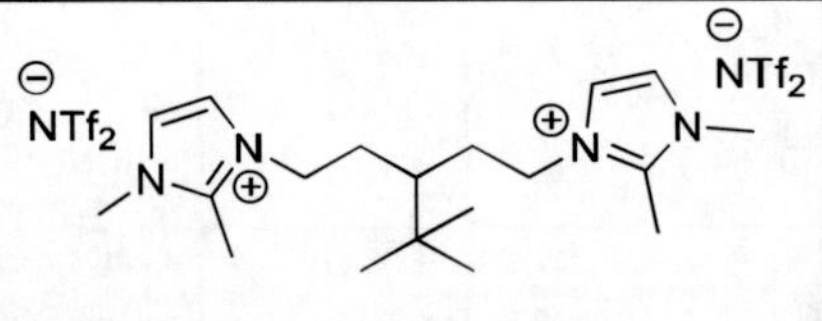 NTf_2^- NTf_2^-	455[a] ($T_{d\ 15\%}$)	(Talebi, Patil, and Armstrong 2018)
36	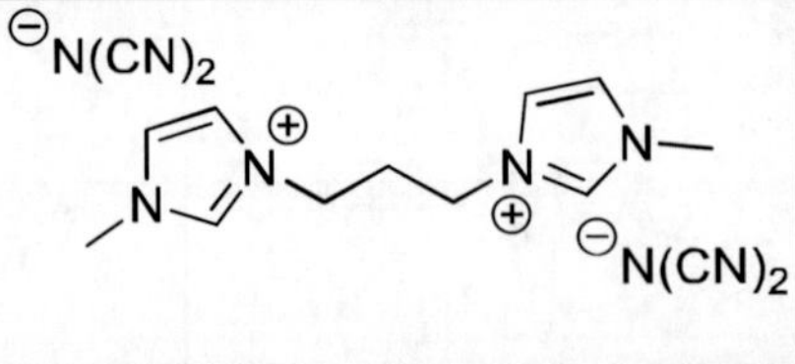 $N(CN)_2^-$ $N(CN)_2^-$	242[a] ($T_{d\ max}$)	(Fareghi-Alamdari et al. 2016)

Entry	Ionic Liquid	T_d (°C)	Reference
37	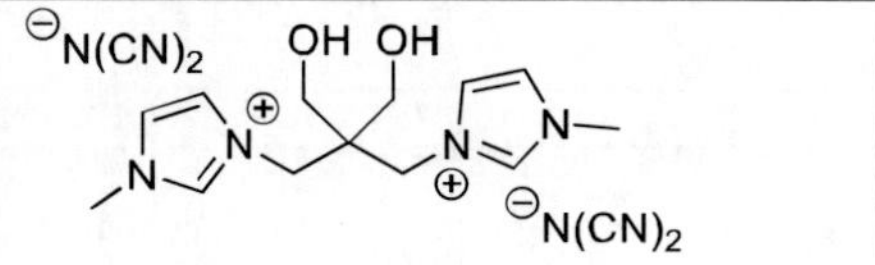 $N(CN)_2$ OH OH N N N N $N(CN)_2$	321[a] ($T_{d\ max}$)	(Fareghi-Alamdari et al. 2016)
38	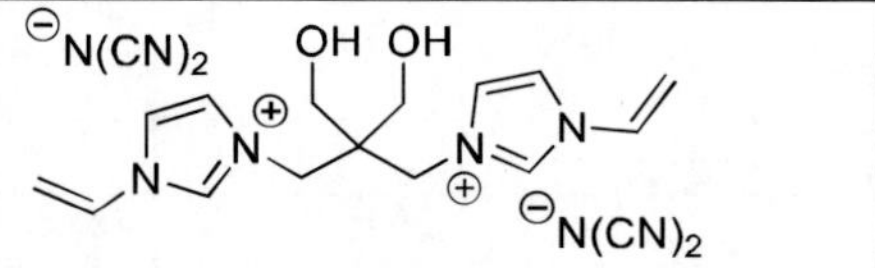 $N(CN)_2$ OH OH N N N N $N(CN)_2$	333[a] ($T_{d\ max}$)	(Fareghi-Alamdari et al. 2016)
39	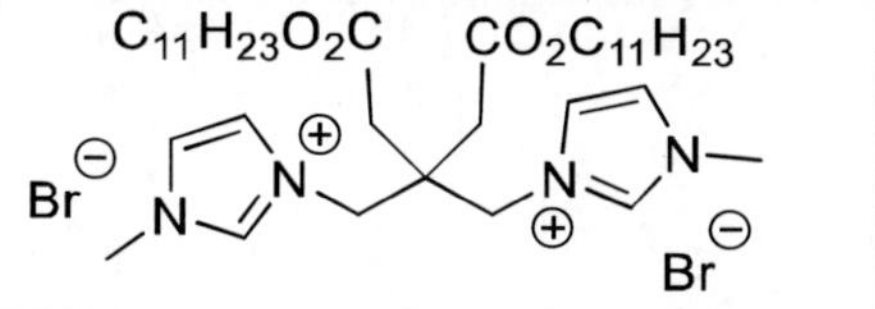 $C_{11}H_{23}O_2C$ $CO_2C_{11}H_{23}$ Br N N N N Br	260[a] ($T_{d\ onset}$)	(Zhuang et al. 2013)
40	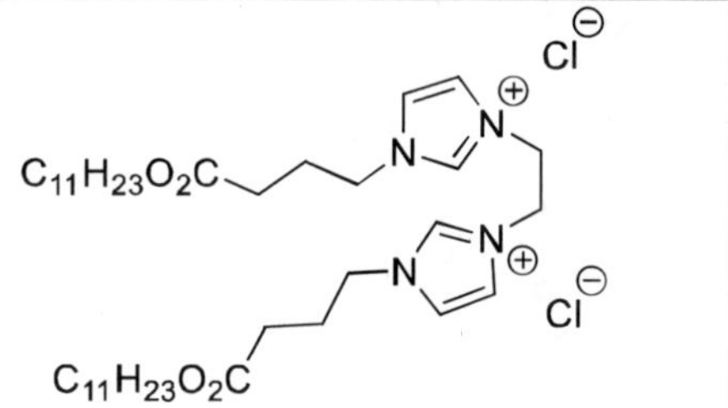 Cl N N $C_{11}H_{23}O_2C$ N N Cl $C_{11}H_{23}O_2C$	291[a] ($T_{d\ onset}$)	(G. Wang et al. 2019)
41	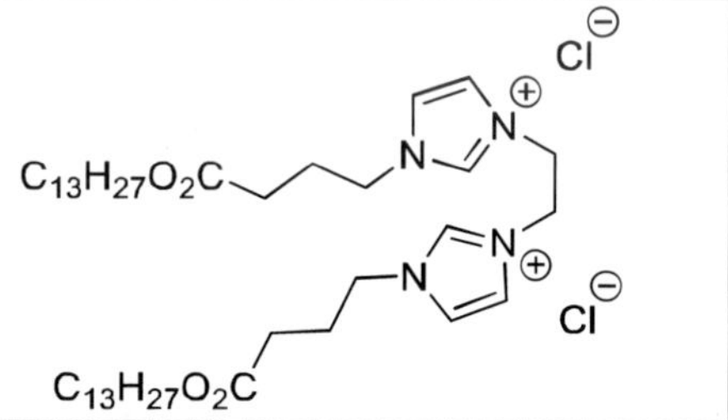 Cl N N $C_{13}H_{27}O_2C$ N N Cl $C_{13}H_{27}O_2C$	279[a] ($T_{d\ onset}$)	(G. Wang et al. 2019)
42	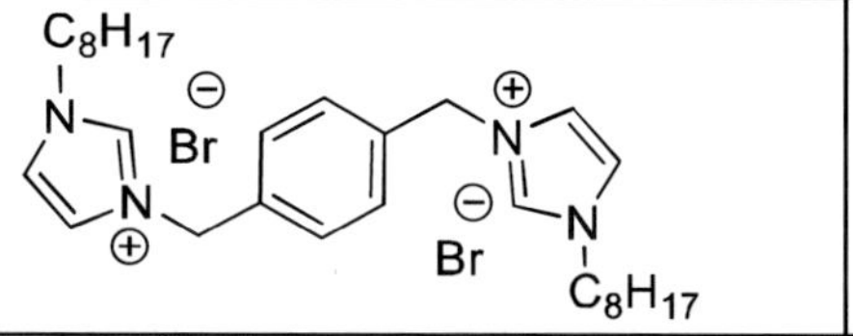 C_8H_{17} N Br N N Br N C_8H_{17}	257[a] ($T_{d\ onset}$)	(Nessim, Zaky, and Deyab 2018)
43	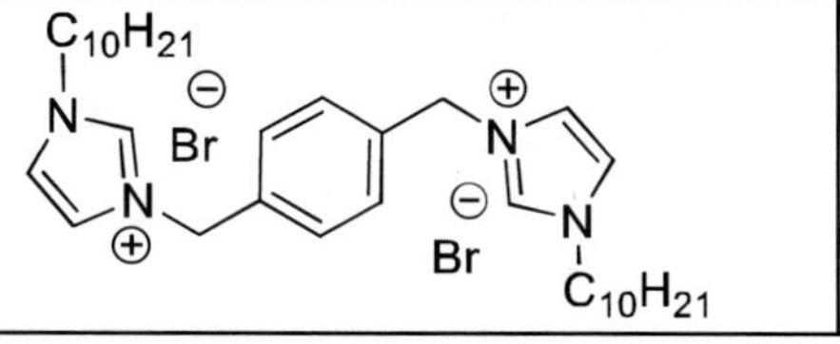 $C_{10}H_{21}$ N Br N N Br N $C_{10}H_{21}$	262[a] ($T_{d\ onset}$)	(Nessim, Zaky, and Deyab 2018)

Table 1. (Continued)

Entry	Ionic Liquid	T_d (°C)	Reference
44	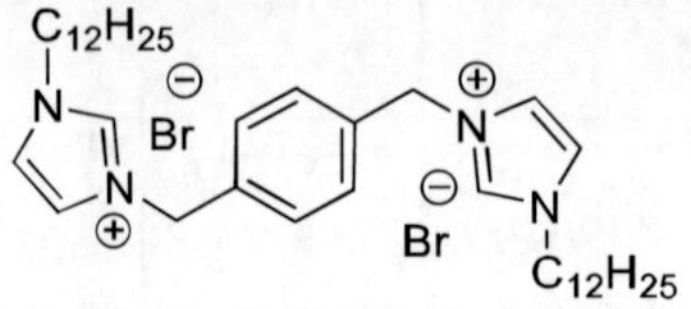	267[a] ($T_{d\ onset}$)	(Nessim, Zaky, and Deyab 2018)
45	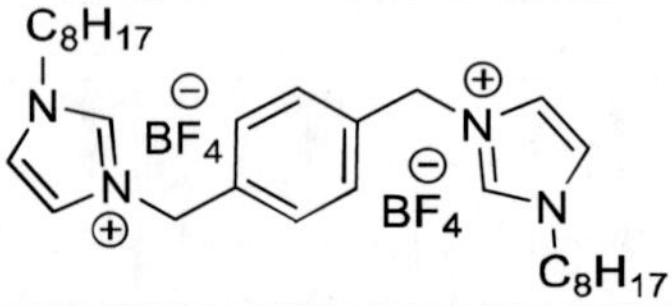	300[a] ($T_{d\ onset}$)	(Zaky, Nessim, and Deyab 2019)
46	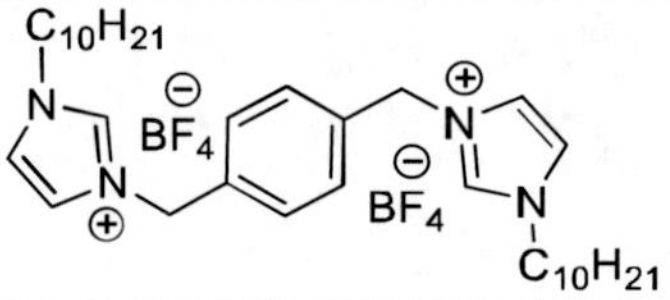	297[a] ($T_{d\ onset}$)	(Zaky, Nessim, and Deyab 2019)
47	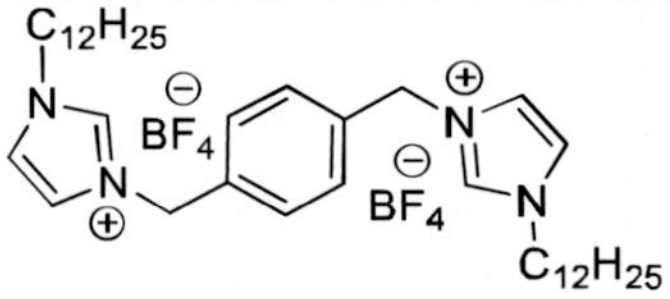	287[a] ($T_{d\ onset}$)	(Zaky, Nessim, and Deyab 2019)
48	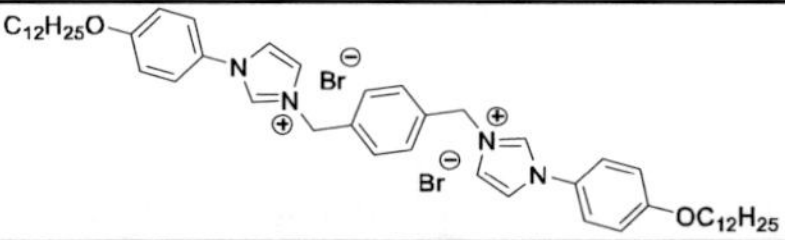	250[a] (T_d)	(Li, Bruce, and Shreeve 2009)
49	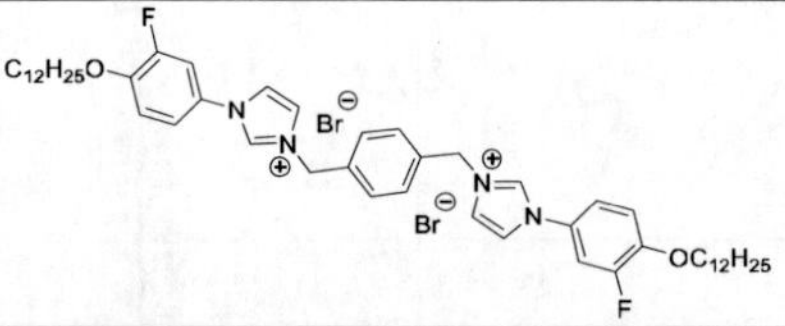	250[a] (T_d)	(Li, Bruce, and Shreeve 2009)
50	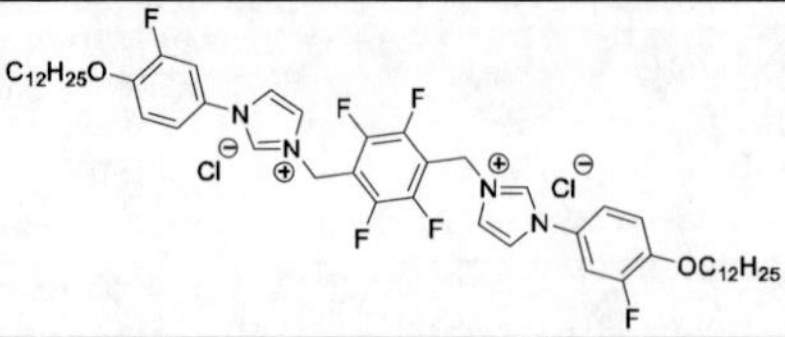	195[a] (T_d)	(Li, Bruce, and Shreeve 2009)
51	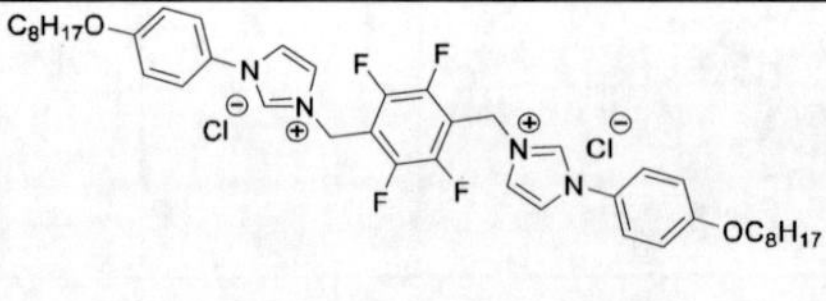	194[a] (T_d)	(Li, Bruce, and Shreeve 2009)

Entry	Ionic Liquid	T_d (°C)	Reference
52	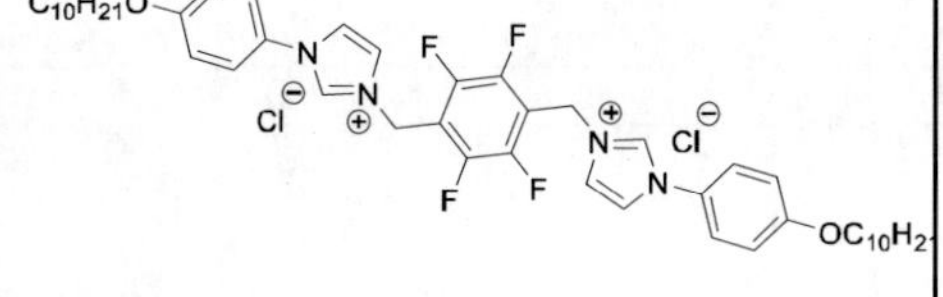	198[a] (T_d)	(Li, Bruce, and Shreeve 2009)
53	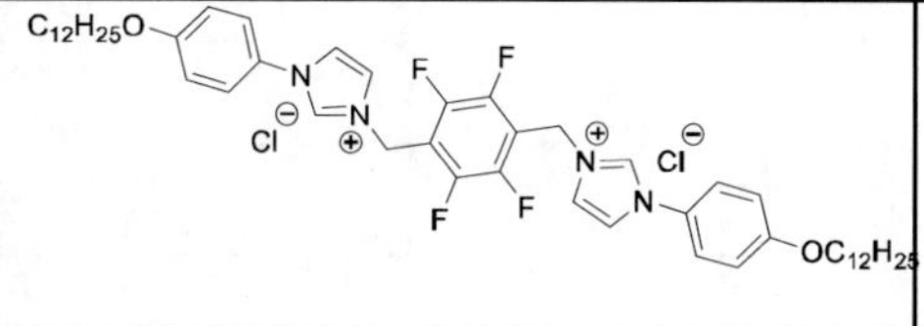	215[a] (T_d)	(Li, Bruce, and Shreeve 2009)
54	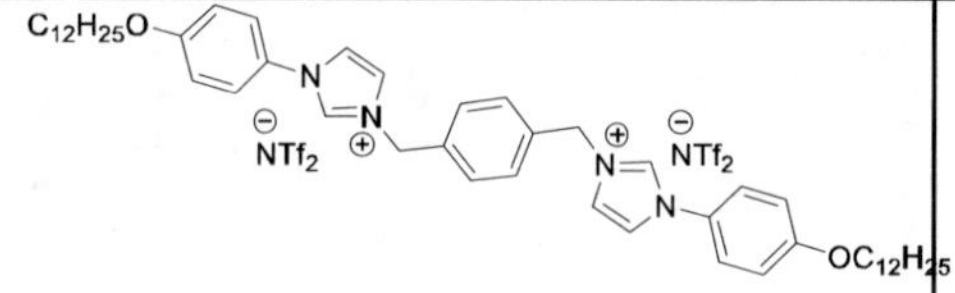	325[a] (T_d)	(Li, Bruce, and Shreeve 2009)
55	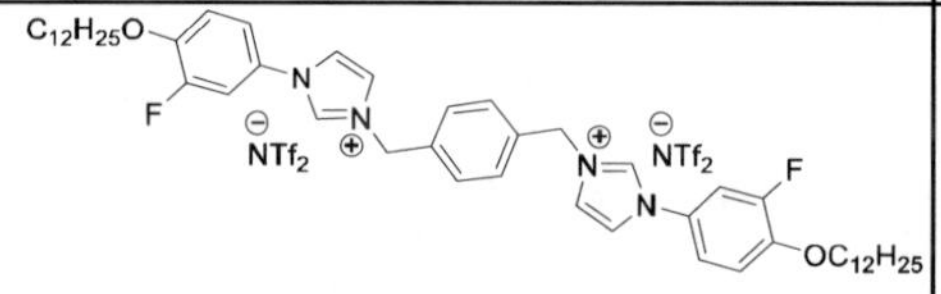	320[a] (T_d)	(Li, Bruce, and Shreeve 2009)
56	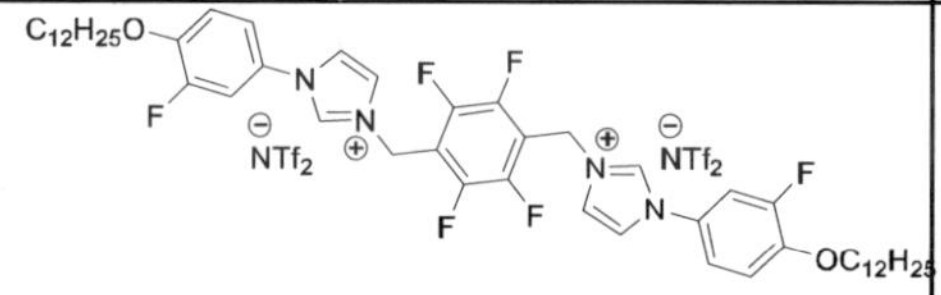	324[a] (T_d)	(Li, Bruce, and Shreeve 2009)
57	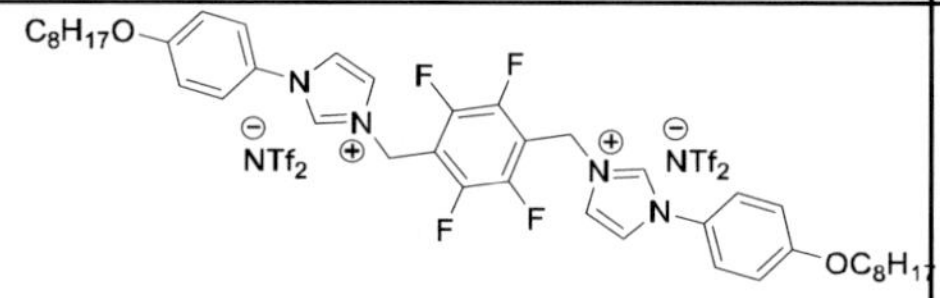	317[a] (T_d)	(Li, Bruce, and Shreeve 2009)
58	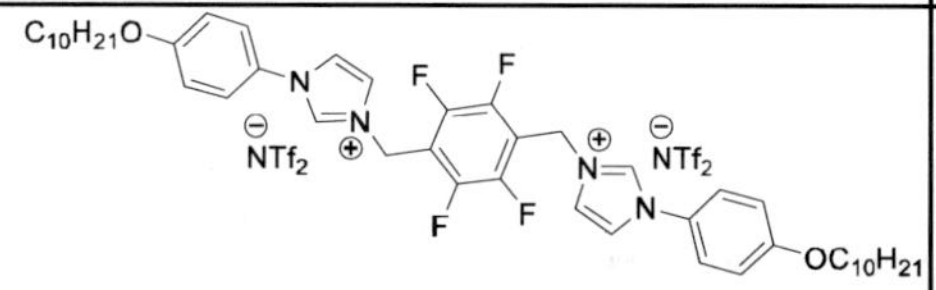	318[a] (T_d)	(Li, Bruce, and Shreeve 2009)
59	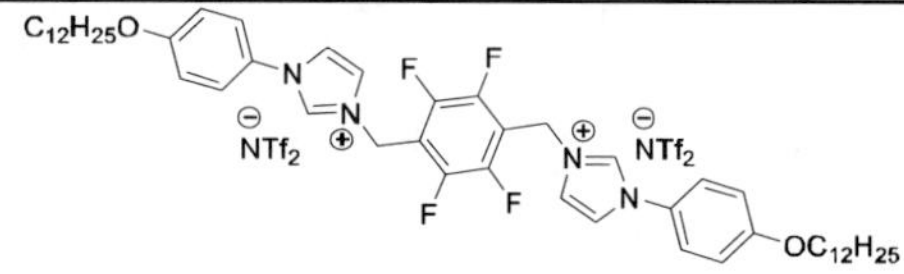	334[a] (T_d)	(Li, Bruce, and Shreeve 2009)

Table 1. (Continued)

Entry	Ionic Liquid	T_d (°C)	Reference
60	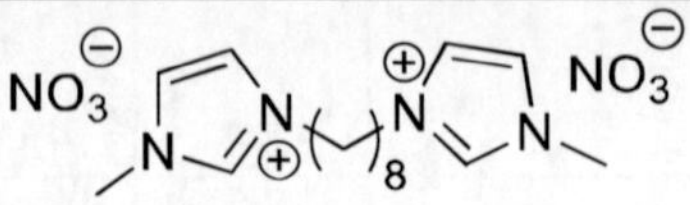 $NO_3^{\ominus}$ $NO_3^{\ominus}$	303[a] ($T_{d\ 10\%}$)	(Shirota et al. 2011)
61	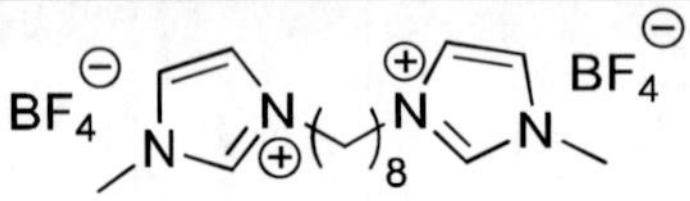 $BF_4^{\ominus}$ $BF_4^{\ominus}$	381[a] ($T_{d\ 10\%}$)	(Shirota et al. 2011)
62	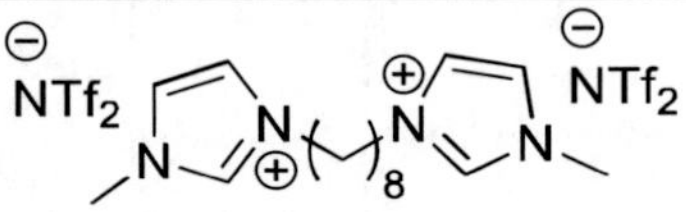 $NTf_2^{\ominus}$ $NTf_2^{\ominus}$	421[a] ($T_{d\ 10\%}$)	(Shirota et al. 2011)
63	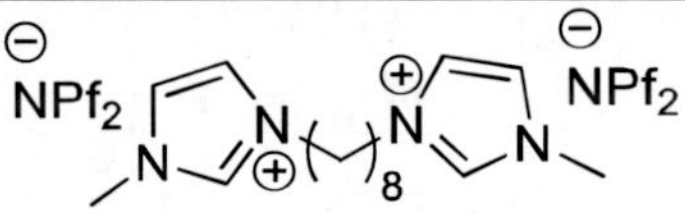 $NPf_2^{\ominus}$ $NPf_2^{\ominus}$	414[a] ($T_{d\ 10\%}$)	(Shirota et al. 2011)
64	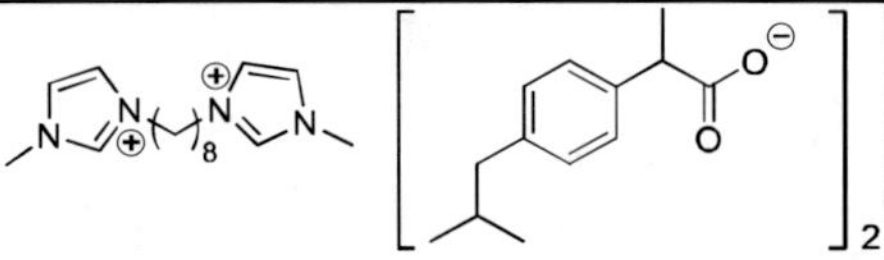	197[a] ($T_{d\ max}$)	(Vieira et al. 2020)
65	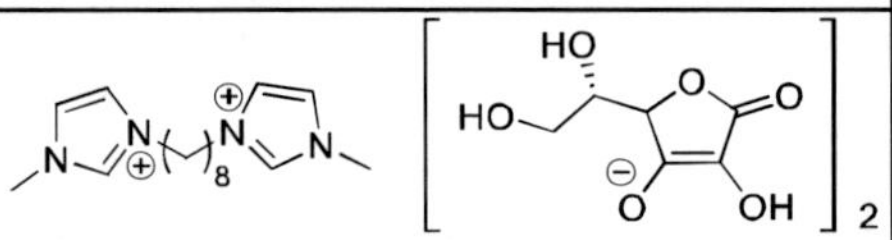	180[a] ($T_{d\ max}$)	(Vieira et al. 2020)
66	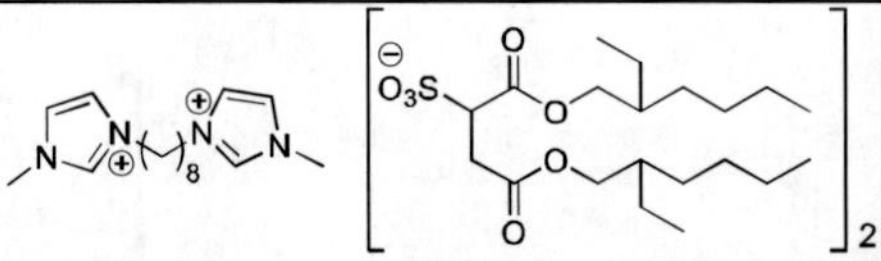	252[a] ($T_{d\ max}$)	(Vieira et al. 2020)
67	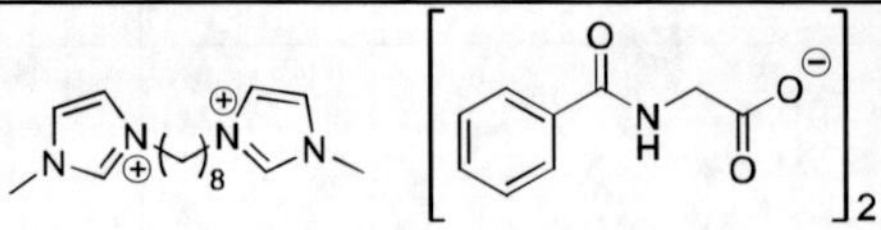	276[a] ($T_{d\ max}$)	(Vieira et al. 2020)
68	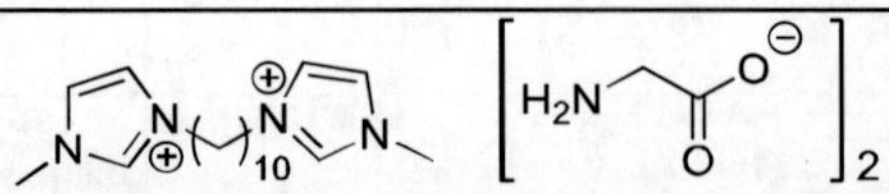	217[a] ($T_{d\ max}$)	(Beck et al. 2022)

Entry	Ionic Liquid	T_d (°C)	Reference
69	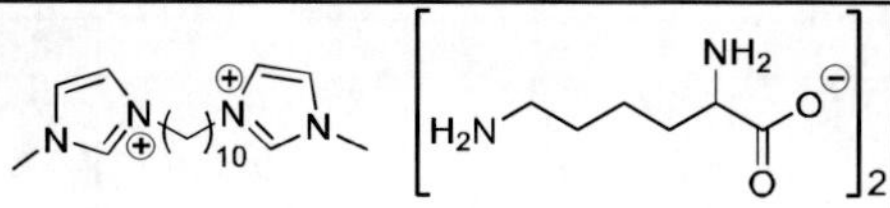 $(CH_2)_{10}$-bridged bis(methylimidazolium) [lysinate]$_2$	283[a] ($T_{d\ max}$)	(Beck et al. 2022)
70	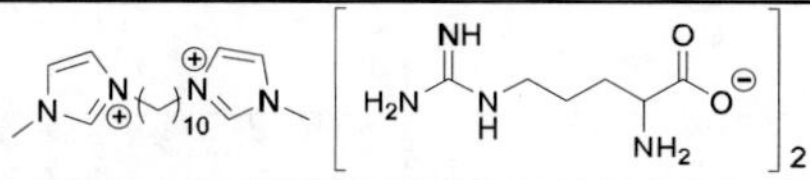 $(CH_2)_{10}$-bridged bis(methylimidazolium) [argininate]$_2$	222[a] ($T_{d\ max}$)	(Beck et al. 2022)
71	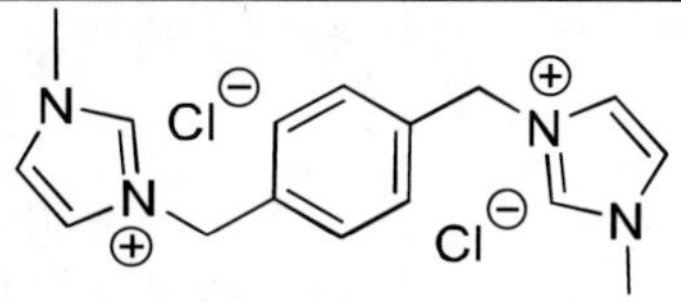 Cl^- / Cl^-	357[a] ($T_{d\ final}$)	(Boumediene et al. 2019)
72	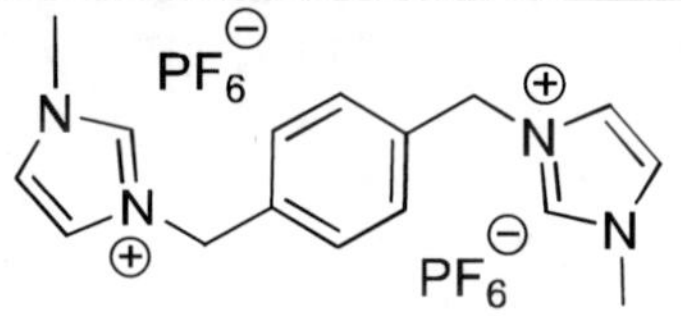 PF_6^- / PF_6^-	476[c] ($T_{d\ final}$)	(Boumediene et al. 2019)
73	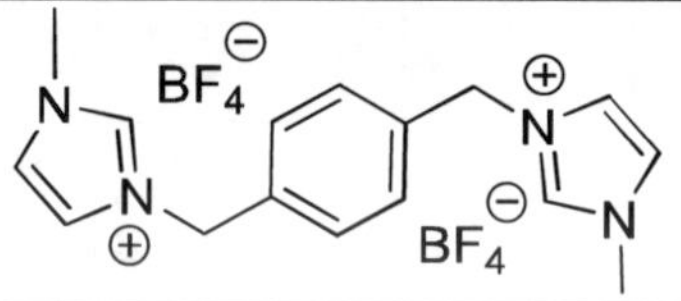 BF_4^- / BF_4^-	407[c] ($T_{d\ final}$)	(Boumediene et al. 2019)
74	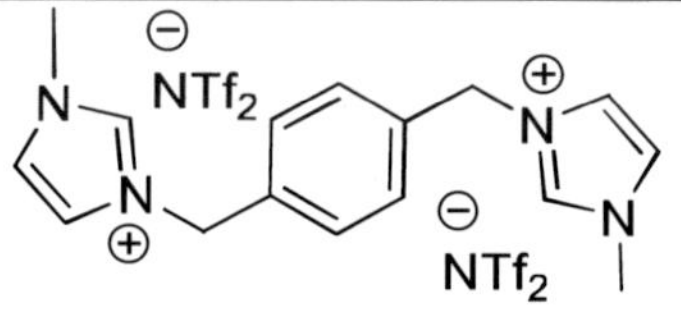 NTf_2^- / NTf_2^-	491[c] ($T_{d\ final}$)	(Boumediene et al. 2019)
75	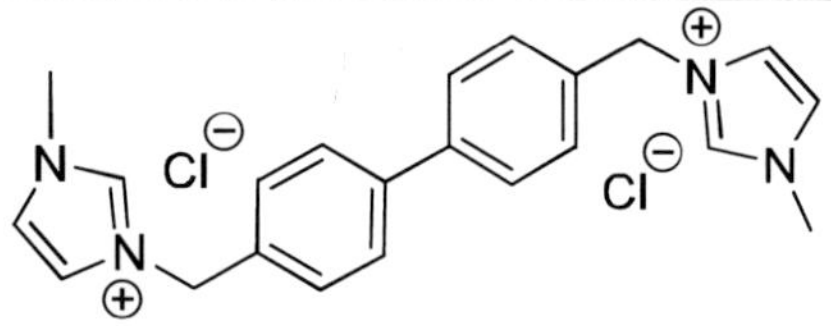 Cl^- / Cl^-	252[c] ($T_{d\ onset}$)	(Boumediene et al. 2020)
76	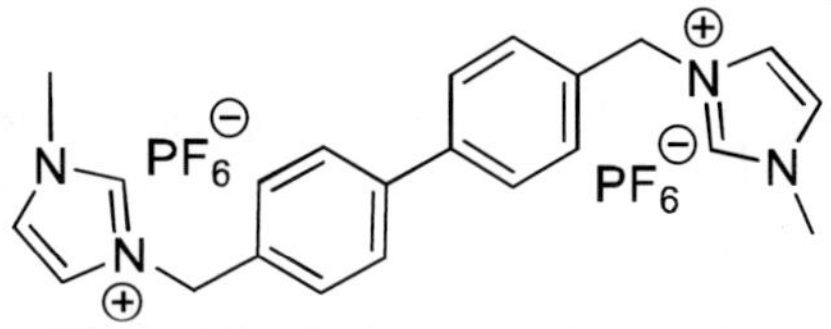 PF_6^- / PF_6^-	323[c] ($T_{d\ onset}$)	(Boumediene et al. 2020)

Table 1. (Continued)

Entry	Ionic Liquid	T_d (°C)	Reference
77	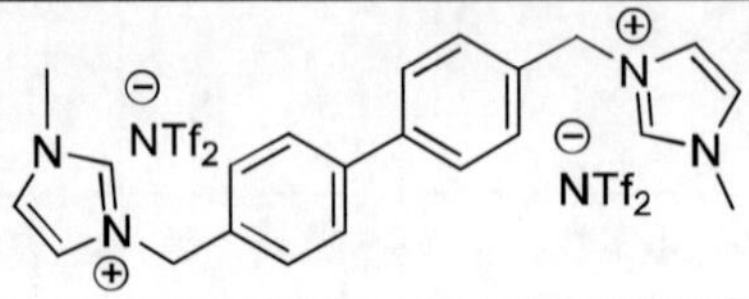	420[c] ($T_{d\ onset}$)	(Boumedienne et al. 2020)
78	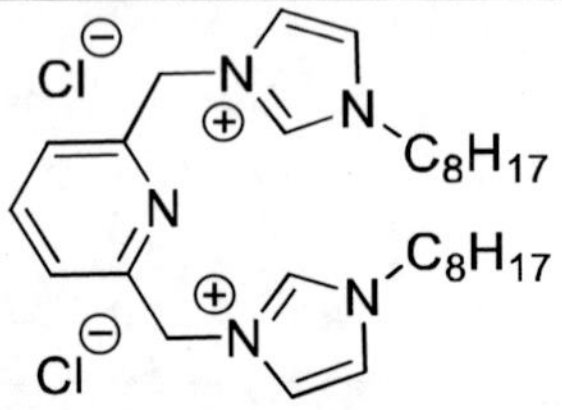	273[a] ($T_{d\ onset}$)	(Clarke et al. 2020)
79	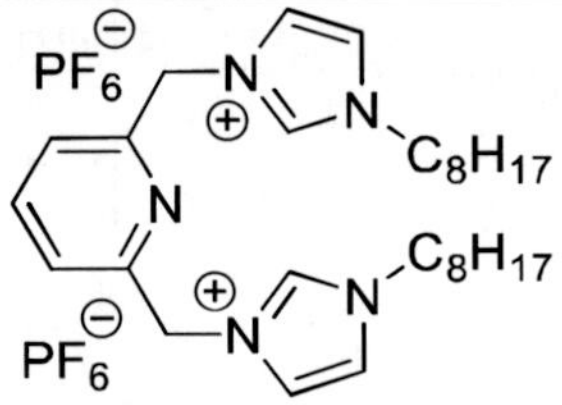	429[a] ($T_{d\ onset}$)	(Clarke et al. 2020)
80	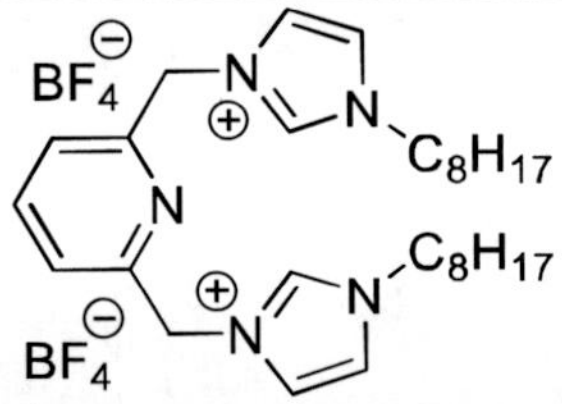	419[a] ($T_{d\ onset}$)	(Clarke et al. 2020)
81	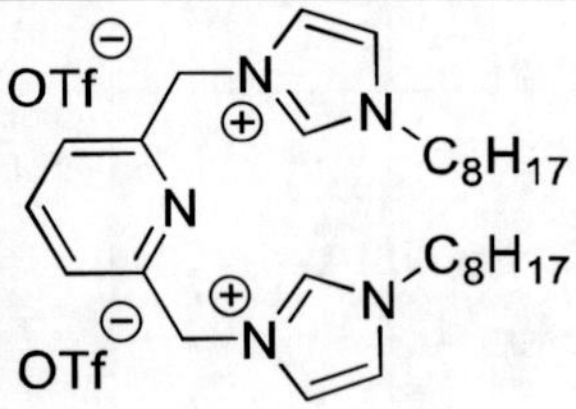	418[a] ($T_{d\ onset}$)	(Clarke et al. 2020)
82	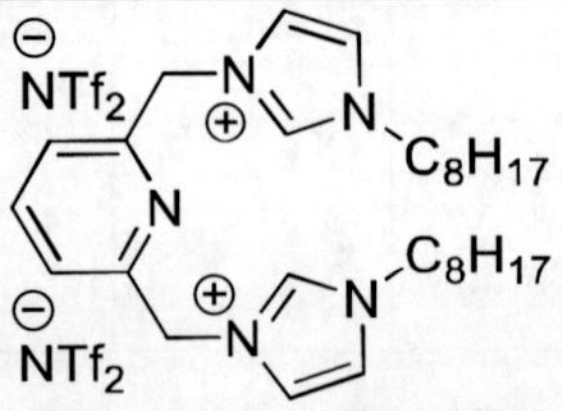	444[a] ($T_{d\ onset}$)	(Clarke et al. 2020)

Entry	Ionic Liquid	T_d (°C)	Reference
83		235[a] ($T_{d\ max}$)	(D'Anna et al. 2013)
84		283[a] ($T_{d\ max}$)	(D'Anna et al. 2013)
85		284[a] ($T_{d\ max}$)	(D'Anna et al. 2013)
86		290[a] ($T_{d\ max}$)	(D'Anna et al. 2013)
87		236[a] ($T_{d\ max}$)	(D'Anna et al. 2013)
88		250[a] ($T_{d\ max}$)	(D'Anna et al. 2013)
89		246[a] ($T_{d\ max}$)	(D'Anna et al. 2013)

Table 1. (Continued)

Entry	Ionic Liquid	T_d (°C)	Reference
90	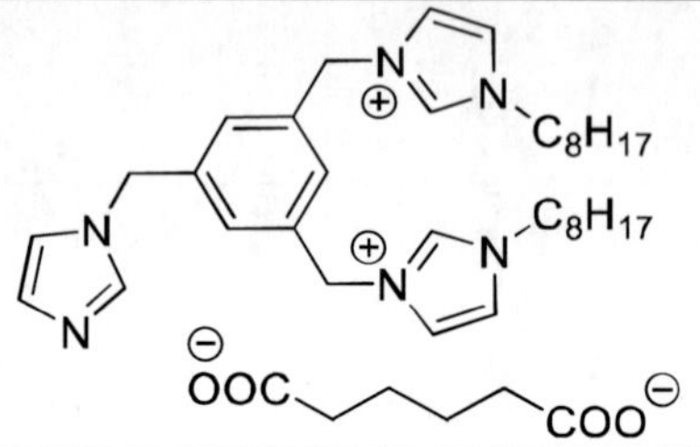	229[a] ($T_{d\,max}$)	(D'Anna et al. 2013)
91	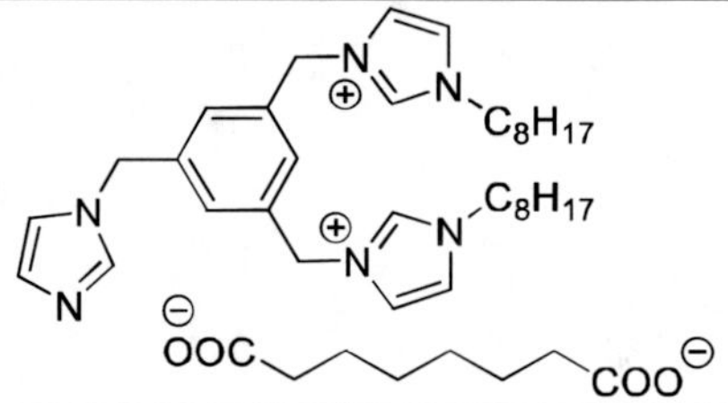	234[a] ($T_{d\,max}$)	(D'Anna et al. 2013)
92	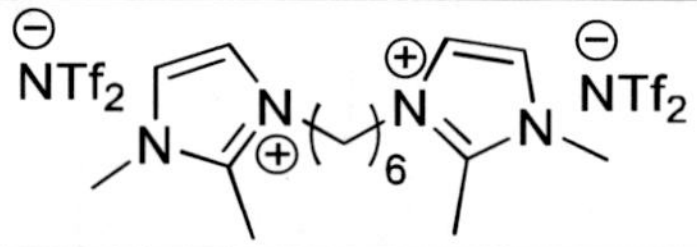	437[a] ($T_{d\,5\%}$)	(Patil et al. 2016)
93	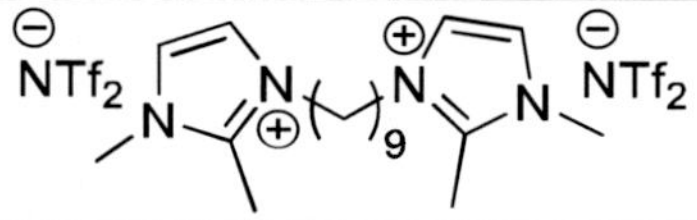	467[a] ($T_{d\,5\%}$)	(Patil et al. 2016)
94	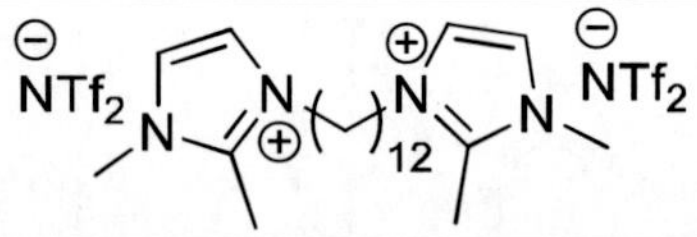	442[a] ($T_{d\,5\%}$)	(Patil et al. 2016)
95	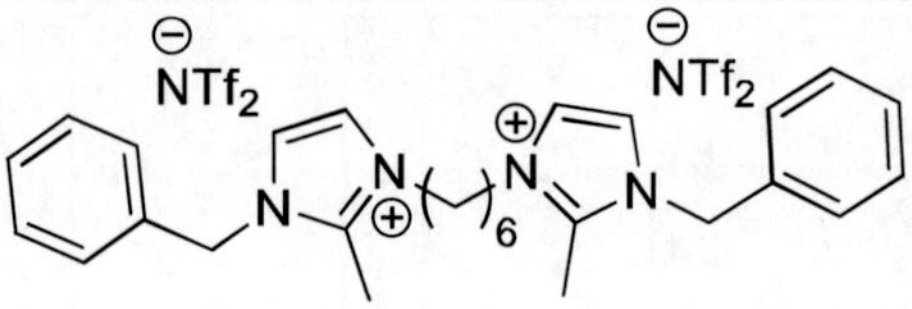	412[a] ($T_{d\,5\%}$)	(Patil et al. 2016)
96	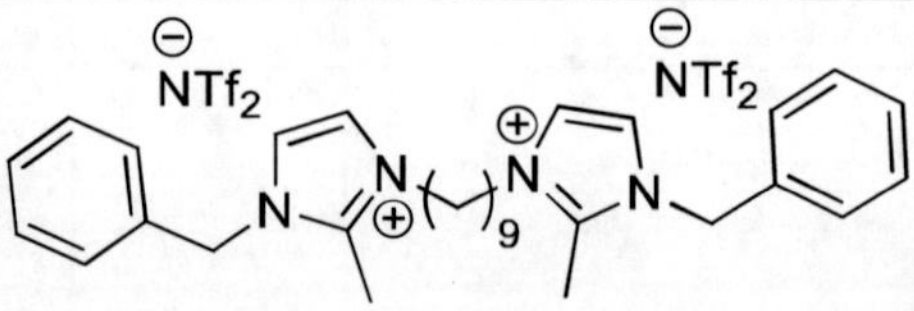	437[a] ($T_{d\,5\%}$)	(Patil et al. 2016)

Entry	Ionic Liquid	T_d (°C)	Reference
97	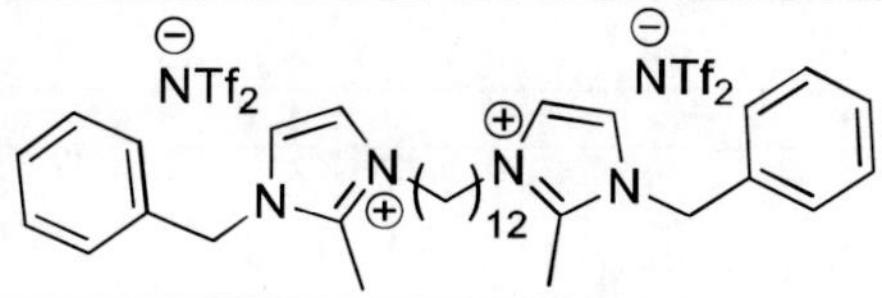 $\ominus$ NTf_2 $\ominus$ NTf_2 N N$\oplus$ $\oplus$N N (12)	394[a] ($T_{d\,5\%}$)	(Patil et al. 2016)
98	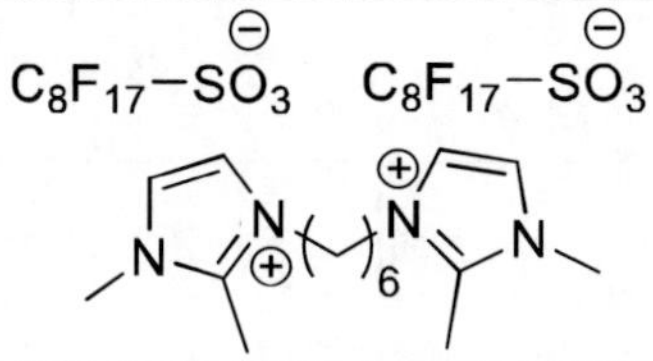 $C_8F_{17}-SO_3^{\ominus}$ $C_8F_{17}-SO_3^{\ominus}$ N N$\oplus$ $\oplus$N N (6)	425[a] ($T_{d\,5\%}$)	(Patil et al. 2016)
99	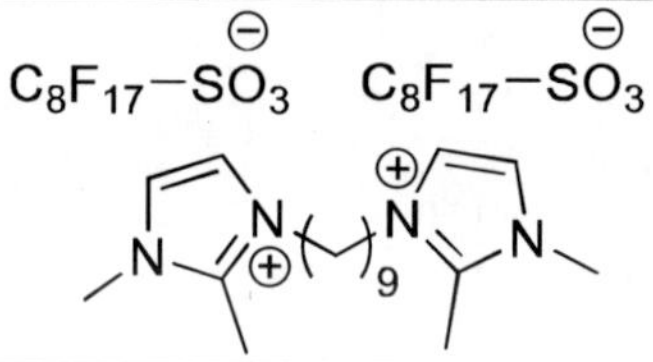 $C_8F_{17}-SO_3^{\ominus}$ $C_8F_{17}-SO_3^{\ominus}$ N N$\oplus$ $\oplus$N N (9)	450[a] ($T_{d\,5\%}$)	(Patil et al. 2016)
100	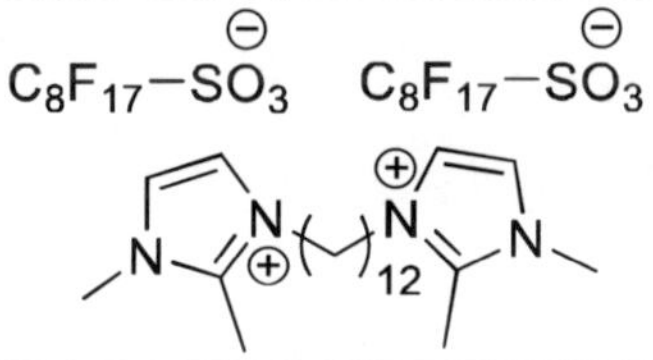 $C_8F_{17}-SO_3^{\ominus}$ $C_8F_{17}-SO_3^{\ominus}$ N N$\oplus$ $\oplus$N N (12)	412[a] ($T_{d\,5\%}$)	(Patil et al. 2016)
101	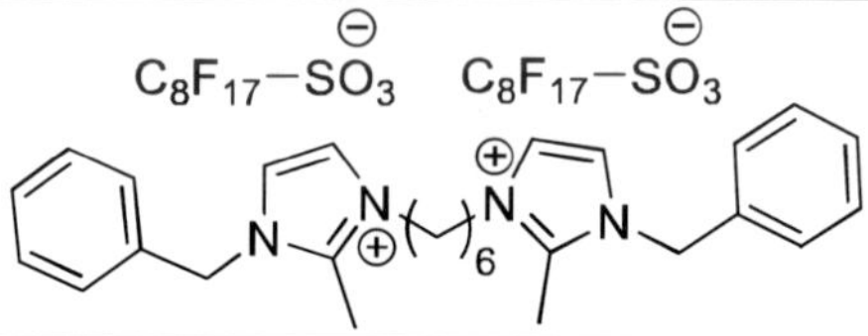 $C_8F_{17}-SO_3^{\ominus}$ $C_8F_{17}-SO_3^{\ominus}$ N N$\oplus$ $\oplus$N N (6)	405[a] ($T_{d\,5\%}$)	(Patil et al. 2016)
102	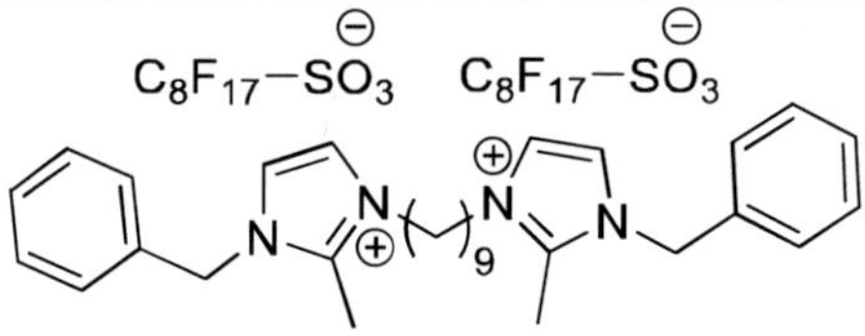 $C_8F_{17}-SO_3^{\ominus}$ $C_8F_{17}-SO_3^{\ominus}$ N N$\oplus$ $\oplus$N N (9)	416[a] ($T_{d\,5\%}$)	(Patil et al. 2016)
103	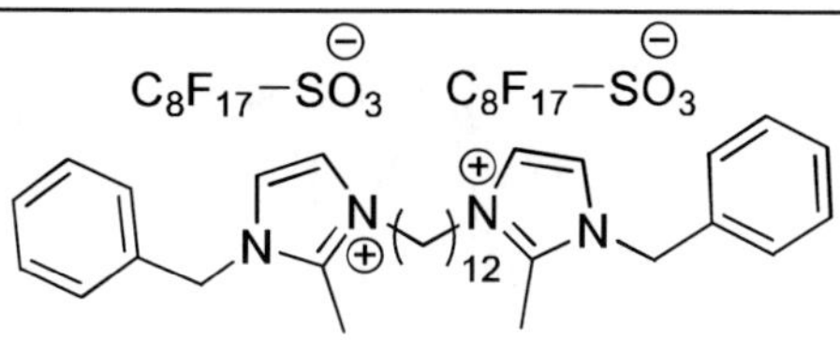 $C_8F_{17}-SO_3^{\ominus}$ $C_8F_{17}-SO_3^{\ominus}$ N N$\oplus$ $\oplus$N N (12)	385[a] ($T_{d\,5\%}$)	(Patil et al. 2016)

Table 1. (Continued)

Entry	Ionic Liquid	T_d (°C)	Reference
104	$[F_3C{-}SO_3]^{-}_2$	332[c] ($T_{d\,5\%}$)	(Talebi et al. 2018)
105	$[C_4F_9{-}SO_3]^{-}_2$	340[c] ($T_{d\,5\%}$)	(Talebi et al. 2018)
106	$[F{-}SO_2{-}N^{-}{-}SO_2{-}F]_2$	232[c] ($T_{d\,5\%}$)	(Talebi et al. 2018)
107	$[F{-}SO_2{-}N^{-}{-}SO_2{-}CF_3]_2$	307[c] ($T_{d\,5\%}$)	(Talebi et al. 2018)
108	$[F_3C{-}SO_2{-}N^{-}{-}SO_2{-}CF_3]_2$	388[c] ($T_{d\,5\%}$)	(Talebi et al. 2018)
109	$[C_2F_5{-}SO_2{-}N^{-}{-}SO_2{-}C_2F_5]_2$	378[c] ($T_{d\,5\%}$)	(Talebi et al. 2018)
110	$[C^{-}(SO_2CF_3)_3]_2$	375[c] ($T_{d\,5\%}$)	(Talebi et al. 2018)
111	$[F_3C{-}SO_3]^{-}_2$	327[c] ($T_{d\,5\%}$)	(Talebi et al. 2018)

Entry	Ionic Liquid	T_d (°C)	Reference
112	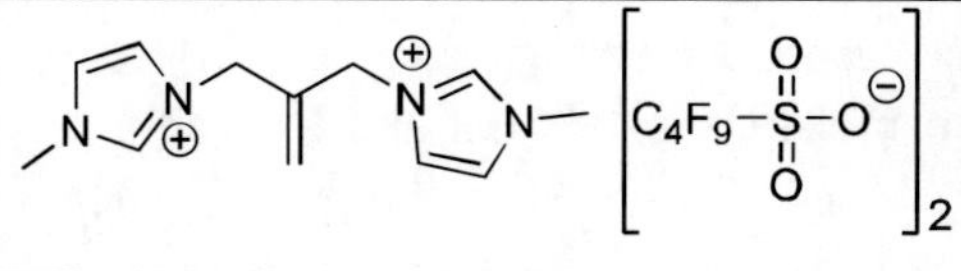	336[c] ($T_{d\,5\%}$)	(Talebi et al. 2018)
113	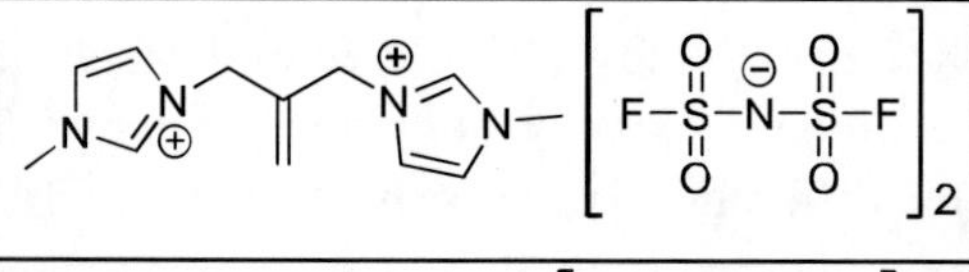	231[c] ($T_{d\,5\%}$)	(Talebi et al. 2018)
114	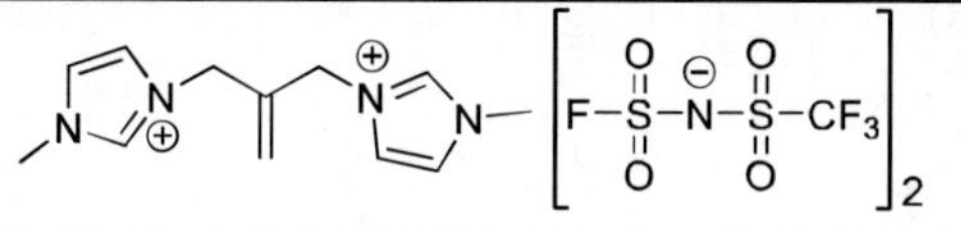	309[c] ($T_{d\,5\%}$)	(Talebi et al. 2018)
115	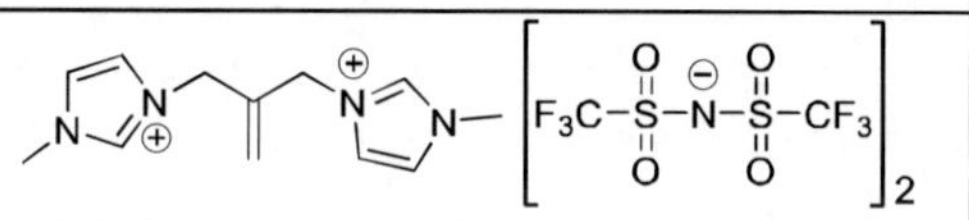	389[c] ($T_{d\,5\%}$)	(Talebi et al. 2018)
116	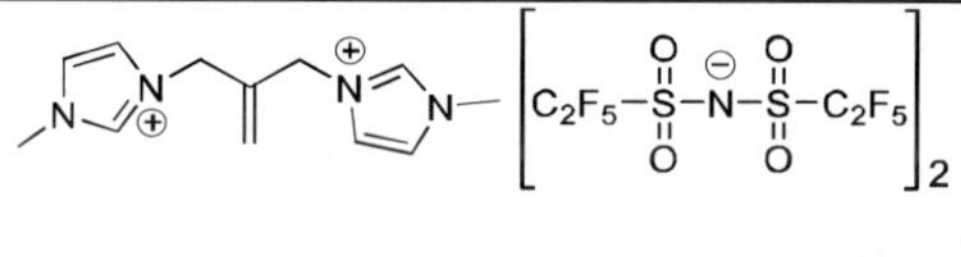	381[c] ($T_{d\,5\%}$)	(Talebi et al. 2018)
117	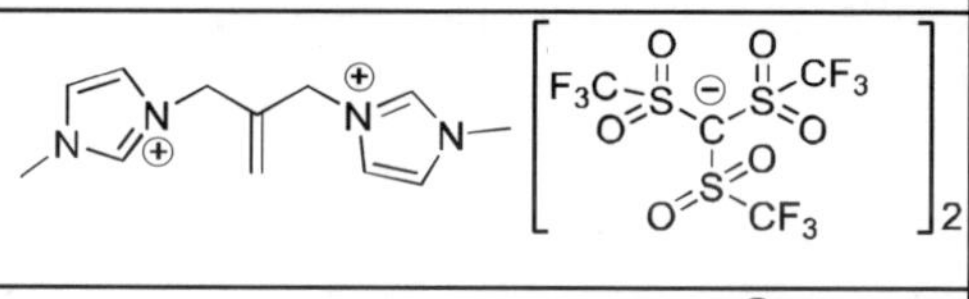	374[c] ($T_{d\,5\%}$)	(Talebi et al. 2018)
118	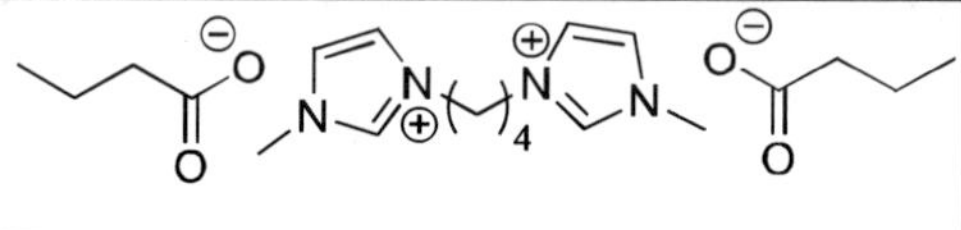	184[a] ($T_{d\,1\%}$)	(Vieira, Villetti, and Frizzo 2021)
119	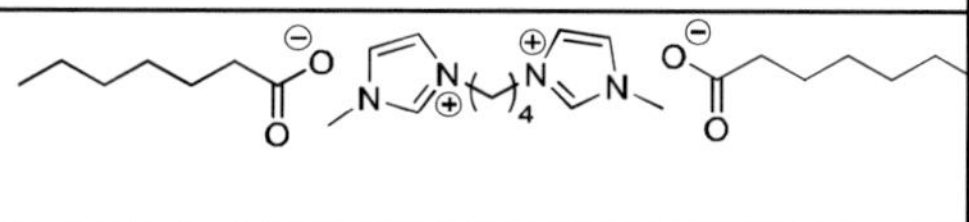	198[a] ($T_{d\,1\%}$)	(Vieira, Villetti, and Frizzo 2021)
120	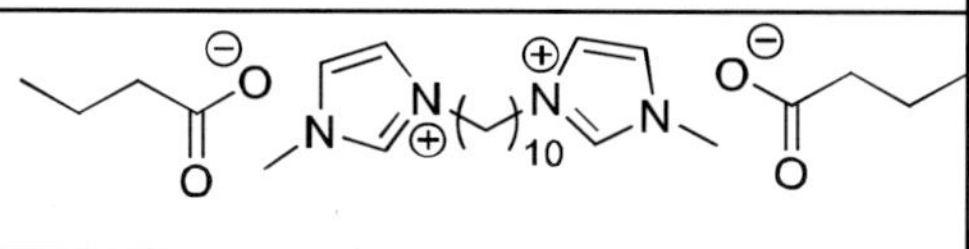	184[a] ($T_{d\,1\%}$)	(Vieira, Villetti, and Frizzo 2021)
121	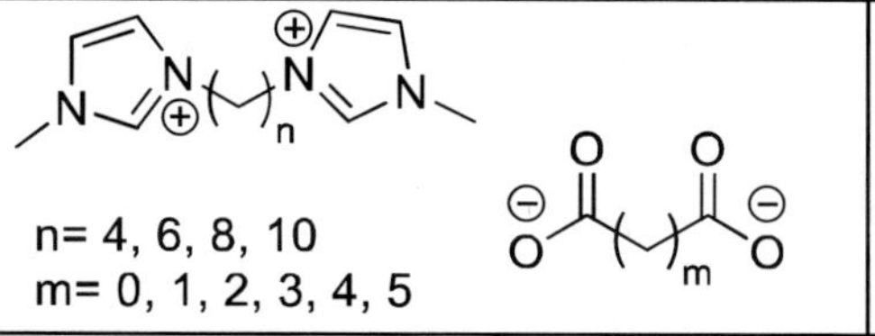 n= 4, 6, 8, 10 m= 0, 1, 2, 3, 4, 5	207-249[a] ($T_{d\,10\%}$)	(Kuhn et al. 2020)

[a]Heating rate of 10°C min^{-1}; [b]Heating rate of 20°C min^{-1}; [c]Heating rate of 5°Cmin^{-1}.

Regardless of the extensive data reported on the decomposition of ILs, it is difficult to make direct comparisons owing to the different parameters utilized by the authors (Kosmulski, Gustafsson, and Rosenholm 2004). This non-standardization may induce erroneous assumptions when comparing different reports of different ILs. In many cases, it is only acceptable to compare ILs discussed in the same study owing to the identical parameters used to obtain the data. Different decomposition temperatures have been reported, such as $T_{dn\%}$=, for different steps of mass loss in the TGA curve, which cannot be accurately compared for different ILs if were not identical. Another parameter that varies from author to author is the heating rate, which also makes the degradation temperatures of different ILs incomparable if the rate differs. Figure 1 illustrates an example of different heating rates for the same compound. A reduction in the heating rate tends to shift the decomposition profile to lower temperatures.

The effect of the heating rate observed in Figure 1 can be explained by the heat-transfer limitation. At low heating rates, a longer time is required for the system to achieve equilibrium, whereas a larger instantaneous thermal energy is provided to the system. At high heating rates for the same time and temperature, the temperature needed to decompose the sample is higher owing to its short reaction time. Thus, with an increase in heating rate, the TGA curve shifted to higher temperatures (Ullah et al. 2016; Slopiecka, Bartocci, and Fantozzi 2012). Other parameters, such as initial sample mass and gas flow (usually nitrogen), may interfere with the TGA curve, but the differences are much more subtle; only extreme differences compromise the data comparison. Even the composition of the pans that hold the sample can modify the results (Siedlecka et al. 2011).

Given the divergences in the reported data, it is important to perform comparisons wisely and to carefully select the studies that are truly comparable. Many structural tendencies have already been traced for imidazolium-based ILs, and we will comment on them in this section, along with the data compiled in Table 1.

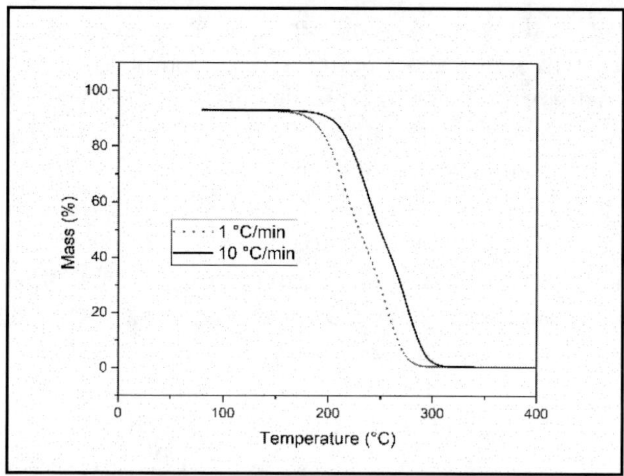

Figure 1. Effect of heating rate on the course of a TGA scan.

Before introducing the structural tendencies that affect the thermal stability of ILs, it is important to remember that the target of this section is dicationic ionic liquids (DILs), since they are less discussed in the literature than their monocationic analogs. However, DILs are interesting because they provide more structural possibilities and are more thermally stable than monocationic ILs (Shirota et al. 2011). The higher thermal stability of DILs is attributed to the higher liquid density owing to the compressibility caused by the addition of a second cationic moiety, which can cause more intermolecular interactions (Shirota et al. 2011). As shown in Table 1, the DIL with a spacer chain length of two methylene groups (entry 2) was approximately 20 °C more stable than its monocationic analog with the same anion (bromide) (entry 1).

DILs are primarily modified through the alteration of their alkyl chain length, either the spacer or the side chain (e.g., entries 2-10, 40-41 and 92-103). An increase in the alkyl chain length decreases the thermal stability of the IL. Increasing the chain length increases the molecular mass and steric hindrance; therefore, it is possible that the stability decreases as the number of alkyl groups attached to the nitrogen increases (Zhang et al. 2019; Claros et al. 2010; Bender et al. 2019). Increasing only one alkyl side-chain length also decreases the thermal stability, and the asymmetry leads to a less organized packing system, making it difficult to organize a crystalline system. The asymmetrical structure in Table 1 is less stable than both symmetrical analogs (i.e., entry 14 compared to entry 2 (Zhang et al. 2019; Claros et al. 2010; Bender et al. 2019)). The methyl group at position 2 (entry 17) of the imidazolium ring underwent no significant change in stability compared with the analogous group at position 3 (entry 4). The addition of a second methyl group seemed to increase the thermal stability (entries 18 and 19); this phenomenon was related to certain decomposition mechanisms (J. Liu et al. 2020) that will be discussed later in this chapter.

The substitution of linear alkyl side-chains with functional groups increases the thermal stability. Functional groups such as those in entries 20–23 of Table 1 increase the thermal stability by their ability to form more intermolecular interactions, which demand more energy to break and thus increases the overall thermal stability (Zhang et al. 2018). Insertion of substituents on the alkyl spacer chain is also possible (entries 24–35), which all lead to a decrease in thermal stability when compared to the non-substituted analog owing to the increase in steric hindrance and the decrease in symmetry (Talebi, Patil, and Armstrong 2018). The decrease in symmetry and increase in steric hindrance result in lower molecular organization, which leads to a decrease in thermal stability.

Similarly to the alkyl side-chains, functional groups such as diols in alkyl spacer chains (entries 37 and 38) increase the thermal stability by forming hydrogen bonds, thereby increasing the stability of the ILs (entry 39 presents a not comparable structure) (Fareghi-Alamdari et al. 2016). Benzylic spacers are also possible, such as the structures shown in entries to 42–44. The decomposition temperatures of these species are quite similar, but the ILs with the longer side-chain length possessed slightly higher decomposition temperatures, which may indicate the role of weak Van der Waals interactions between alkyl chains in the thermal stability (Nessim, Zaky, and Deyab 2018). Even with a different anion (e.g., tetrafluoroborate), this tendency remained the same (entries 45-47) (Zaky, Nessim, and Deyab 2019).

Large cationic structures are minimally affected by structural changes, such as the polyaromatic structures in entries 48–59. Modification of the cation induces only minor changes in the thermal stability and does not follow a clear trend (Li, Bruce, and Shreeve 2009). While the influence on the cation is more subtle, the anion structure drastically affects the thermal profile (Siedlecka et al. 2011). Among inorganic anions, ILs of halides such as chloride and bromide are the least thermally stable. This stability is attributed to the coordination ability of the anions. Halides are much more strongly coordinating than other inorganic anions, such as hexafluorophosphate (PF_6^-) and tetrafluoroborate (BF_4^-), which increase the thermal stability of ILs, along with anions containing sulfonyl groups (Siedlecka et al. 2011).

The thermal stability attributed to the anionic species of ILs is usually related to their nucleophilicity (less nucleophilic, higher thermal stability), size (larger anions result in higher thermal stability), polarizability (more polarized anions interact more strongly with the cation, resulting in higher thermal stability), and the presence of fluorinated groups (anions containing fluorine in the structure are usually more thermally stable). Based on literature reports, the thermal stability of ILs containing the following anions is as follows: Cl^- < BF_4^- < PF_6^- < NTf_2^-. ILs containing bis(trifluoromethylsulfonyl)imide (NTf_2^-) are among the most thermally stable ILs known, and ILs containing halide anions such as chloride, as well the triflate anion ($CF_3SO_3^-$ or TfO^-), are among the least stable (Boumediene et al. 2019; Shirota et al. 2011; Clarke et al. 2020). Examples of each are given in entries 60–63,71–74, 75–77, and 78–82.

A comparison of multiple fluorinated anions showed that NTf_2^-, NPf_2^- (bis(perfluoroethylsulfonyl)imide), and CTf_3^- (tris(trifluoromethanesulfonyl) methide) yielded the most thermally stable ILs, while ILs containing TfO^- and NfO^- ($C_4F_9SO_3^-$) were the least stable, which is attributed to the significant nucleophilicity of these anions compared with the rest of the series (Talebi et al. 2018).

Organic anions usually result in less thermally stable ILs compared to those composed of inorganic anions. Some aspects remain the same for organic anions; for example, more coordinating or more nucleophilic anions lead to less thermally stable ILs. Since the conjugated base (anion) of a strong acid is more stable than that of a weak acid, it can also be assumed that anions derived from strong acids increase the thermal stability of ILs, while those from weak acids decrease the thermal stability (Siedlecka et al. 2011; Khan et al. 2017). Although organic anions are more complex than inorganic anions owing to the possibility of intermolecular reactions and the reactivity of the anion itself, some of the structural tendencies of inorganic anions also apply to organic anions. Larger anions along with higher polarizability and minimum steric hindrance impart higher thermal stability to ILs (Vieira et al. 2020; Beck et al. 2022). Examples of organic anions with different structures, such as amino acids, and thus different decomposition temperatures of ILs are shown in entries 64–70 of Table 1. Organic anionic species frequently originate from carboxylic acids, such as aliphatic carboxylic acids. ILs with aliphatic carboxylic acids (entries 118–120) follow the same trend as the spacer or side-chain length of the cations, that is, the longer the alkyl chain length of the carboxylate anion, the lower the thermal decomposition temperature (Vieira, Villetti, and Frizzo 2021).

Despite the fact that carboxylate anions follow the above trend, ILs with dicarboxylate species did not appear to exhibit the same tendency. Dicarboxylate anions with different alkyl spacer chain lengths combined with cations with different alkyl spacer chain lengths have similar decomposition temperatures (entry 121) (Kuhn et al. 2020).

The nature of the spacer link in organic dianions has already been investigated (entries 86–91). The thermal stability seems to be affected by both the nature of the charged moiety and that of the spacer link. An aromatic spacer link seems to contribute to higher thermal stability of the IL owing to stabilization by resonance. This means that dianions with the same anionic moiety and an aliphatic link are thermally stable. Interestingly, an increase in the extension of the aromatic link seemed to decrease the thermal stability (entries 88 and 89). In addition, the species in entries 90 and 91 demonstrated a different tendency: the dicarboxylate with a longer spacer chain was more thermally stable, which may indicate that a more flexible anion is more suitable for this type of structure (D'Anna et al. 2013).

Based on this summary of structural tendencies affecting the thermal stability of DILs, it can be concluded that it is possible to compile such data from the literature to estimate the thermal stability and structural tendencies of these compounds. However, there is still a lack of standardization about the parameters used to determine the thermal stability results, making it impossible to compare studies and make assumptions about the data (Siedlecka et al. 2011; Kosmulski, Gustafsson, and Rosenholm 2004). In addition, it is very important to remember that the comparisons made using degradation parameters such as T_{onset} are no more than short-term thermal stabilities. These values were overestimated by the rapid temperature ramping rate. Even though the short-term thermal stabilities can be used to gauge tendencies and make general comparisons, for real applications of ILs that are directly dependent on the thermal stability, only data obtained by kinetic models can predict the real (long-term) thermal stability of ILs (Wooster et al. 2006; Ullah et al. 2016; Baranyai et al. 2004).

Decomposition Kinetics

To completely elucidate the long-term thermal stability of a given compound, it is important to determine quantitative parameters that allow for confident predictions and analysis. This can be achieved by determining the kinetic parameters, that is, the activation energy (E_a) and pre-exponential factor (A), of the thermal decomposition process. This section aims to provide overall guidelines on how these parameters are calculated and discuss the published studies that use this type of analysis for dicationic imidazolium-based ILs.

Basics of Kinetic Analysis of Thermal Decomposition Processes

The study of the decomposition rates in thermally stimulated processes and subsequent determination of the kinetic parameters (E_a *and* A) are the main focus of the kinetic analysis of thermal decomposition processes. These studies allow us to predict the decomposition rates and acquire information about the decomposition mechanisms. There are two principal methodologies used in the kinetic analysis of TGA data: isothermal methodology, in which samples are maintained at different temperatures for specific time intervals; and isoconversional methodology, where different heating rates are used in the kinetic analysis (S. Vyazovkin 2015).

In both methodologies, the processes are analyzed not in terms of mass but conversion (α), which is calculated from the mass-loss data according to Equation 1, where m_i and m_f are the initial and final masses of the sample, respectively, and m_T is the mass at temperature T. The rate of this decomposition process ($d\alpha/dt$) can be expressed as Equation 2, where $k(T)$ is the rate constant and $f(\alpha)$ is a theoretical kinetic model. The rate constant $k(T)$ is dependent on the temperature according to the Arrhenius equation (Equation 3), where A is the pre-exponential factor, E_a is the activation energy, and R is the universal gas constant. Substituting Equation 3 into Equation 2 yields Equation 4, which relates the decomposition rate, $d\alpha/dt$, to the conversion, α, which is represented by a kinetic model, $f(\alpha)$ (S. Vyazovkin 2015).

$$\alpha = \frac{m_i - m_T}{m_i - m_f} \tag{1}$$

$$\frac{d\alpha}{dt} = k(T)f(\alpha) \tag{2}$$

$$k(T) = Aexp\left(\frac{-E_a}{RT}\right) \tag{3}$$

$$\frac{d\alpha}{dt} = Aexp\left(\frac{-E_a}{RT}\right)f(\alpha) \tag{4}$$

Several *f(α)* depends on the profile of the conversion curve, as shown in Table 2, where *g(α)* is the integrated form of *f(α)*. However, according to Vyazovkin (S. Vyazovkin 2015), it is not necessary to apply all the *f(α)* values in Table 2 to the experimental data to determine if one is the most optimal. As shown in Figure 2, there are three main conversion curve profiles: acceleration, deceleration, and sigmoidal. In the accelerating profile (curve 1 in Figure 2), α increases with temperature (or time) only at the end of the decomposition process, and the kinetic models that best fit this type of data are the power-law models (entries 1–4 in Table 2). In the decelerating case (curve 2 in Figure 2), α rapidly increased at the beginning of the process, and the best fit was obtained by using the kinetic models for diffusion (entries 5, 10, and 13 in Table 2), geometry (entries 11 and 12 in Table 2), or reaction order (entry 6 in Table 2). In the case of a sigmoidal α curve profile (curve 3 in Figure 2), the Avrami–Erofeev model is the optimum kinetic model (entries 7–9 in Table 2). These tendencies help to define an adequate kinetic model; however, it is not uncommon to obtain good fits using different kinetic models for the same set of experimental data (S. Vyazovkin 2015).

Table 2. Examples of kinetic models for thermally stimulated processes

	Kinetic model (code)	*f(α)*	*g(α)*
1	Power Law (P4)	$4\alpha^{3/4}$	$\alpha^{1/4}$
2	Power Law (P3)	$3\alpha^{2/3}$	$\alpha^{1/3}$
3	Power Law (P2)	$2\alpha^{1/2}$	$\alpha^{1/2}$
4	Power Law (P2/3)	$2/3\alpha^{-1/2}$	$\alpha^{3/2}$
5	Unidimensional Diffusion (D1)	$1/2\alpha^{-1}$	α^2
6	Mampel (first order) (F1)	$1 - \alpha$	$-\ln(1 - \alpha)$
7	Avrami–Erofeev (A4)	$4(1 - \alpha)[-\ln(1- \alpha)]^{3/4}$	$[-\ln(1 - \alpha)]^{1/4}$
8	Avrami–Erofeev (A3)	$3(1 - \alpha)[-\ln(1- \alpha)]^{2/3}$	$[-\ln(1 - \alpha)]^{1/3}$
9	Avrami–Erofeev (A2)	$2(1 - \alpha)[-\ln(1- \alpha)]^{1/2}$	$[-\ln(1 - \alpha)]^{1/2}$
10	Tridimensional Diffusion (D3)	$3/2(1 - \alpha)^{2/3}[1 - (1- \alpha)^{1/3}]^{-1}$	$[1 - (1 - \alpha)^{1/3}]^2$
11	Contracting Sphere (R3)	$3(1 - \alpha)^{2/3}$	$1 - (1 - \alpha)^{1/3}$
12	Contracting Cylinder (R2)	$2(1 - \alpha)^{1/2}$	$1 - (1 - \alpha)^{1/2}$
13	Bidimensional Diffusion (D2)	$[-\ln(1 - \alpha)]^{-1}$	$(1 - \alpha)\ln(1 - \alpha) + \alpha$

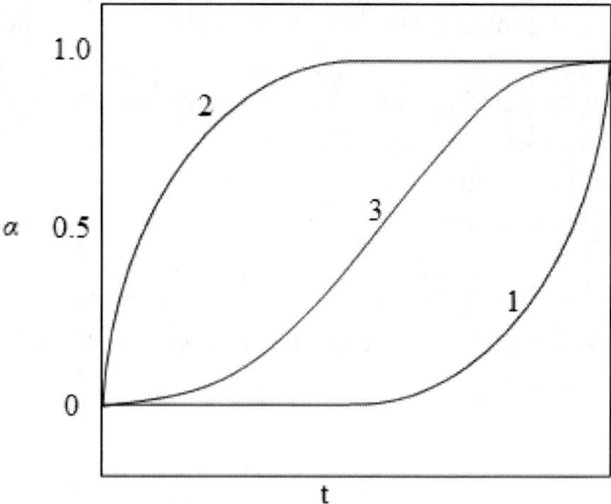

Figure 2. Accelerating (1), decelerating (2), and sigmoidal (3) conversion curves.

Isothermal Kinetic Analysis of Thermal Decomposition Processes

In TGA experiments, it is possible to maintain the sample at a specific temperature and measure mass loss over time. The conversion α is then calculated from Equation 1 by substituting m_T for m_t, which is the mass at time t. The profile of the α vs. t curve (Figure 2) defines the set of $f(\alpha)$ to be used in the kinetic analysis. Since TGA is an integral method (S. Vyazovkin 2015), integrating Equation 4 from zero to α provides Equation 5, which depends on $f(\alpha)$ (see the $g(\alpha)$ expressions in Table 2).

$$g\,(\alpha) \equiv \int_0^{\alpha} \frac{d\alpha}{f(\alpha)} = k(T)t \tag{5}$$

From Equation 5, it can be seen that a plot of $g(\alpha)$ vs. t will have a slope equal to the rate constant $k(T)$. If we obtain the rate constants at a set of different temperatures and apply them to the logarithmic form of Equation 3, yielding Equation 10, the obtained curve will have a slope equal to $-E_a/R$ and an intercept equal to ln$A$ (S. Vyazovkin 2015).

$$\ln k(T) = \ln A - \frac{E_a}{RT} \tag{10}$$

In brief, by performing a set of five isothermal TGA experiments, it is possible to obtain five *k(T)* values; when plotted against $1/T$, the kinetic parameters E_a and A of the isothermal decomposition process are determined. However, this methodology is somewhat problematic because it is necessary to select a theoretical kinetic model to fit the experimental data, leading to unwanted errors in the analysis.

Isoconversional Kinetic Analysis of Thermal Decomposition Processes

This methodology is based on the isoconversional principle, which states that the rate of a certain conversion is a function of temperature only. This principle is demonstrated by the differential logarithmic form of Equation 2 at constant α (Equation 11). At constant α, the second term on the right side of Equation 11 is null, and considering Equation 3, the first term is equal to $-E_a/R$, giving Equation 12 (S. Vyazovkin 2015).

$$\left[\frac{\partial \ln(d\alpha/dt)}{\partial T^{-1}}\right]_\alpha = \left[\frac{\partial \ln k(T)}{\partial T^{-1}}\right]_\alpha + \left[\frac{\partial \ln f(\alpha)}{\partial T^{-1}}\right]_\alpha \tag{11}$$

$$\left[\frac{\partial \ln(d\alpha/dt)}{\partial T^{-1}}\right]_\alpha = -\frac{E_a}{R} \tag{12}$$

Equation 12 indicates that it is possible to obtain the value of E_a for a specific α without having to choose a theoretical kinetic model. This advantage of the isoconversional methodology allows the determination of error-free kinetic parameters, making it possible to calculate $_a$ for each α of the process and offering information about the mechanisms of action. The International Confederation for Thermal Analysis and Calorimetry (ICTAC) considers isoconversional analysis the best methodology for studying the kinetics of any thermally stimulated process (Sergey Vyazovkin et al. 2011; 2014).

To determine the value of E_a for each α, it is necessary to determine the dependence of the isoconversional rate on temperature. This can be achieved by performing a set of TGA experiments using different heating rates (four or five runs are considered sufficient). Several methods can be used to perform these calculations. A constant heating rate, q, expressed in Equation 13, and the variation of α over time, $d\alpha/dt$, can be related to the variation of α over the temperature range, $d\alpha/dT$, using Equation 14. The logarithmic form of Equation 4 can be written in the form of Equation 15, which is used in the Friedman method (Friedman 1964) for the determination of E_a in relation to α. From Equation 15, the slope of the curve $\ln[(d\alpha/dT)q]$ vs. $1/T$ is equal to $-E_a/R$ (S. Vyazovkin 2015). It is noteworthy that the differentiation of TGA data ($d\alpha/dT$) leads to some degree of error in E_a estimations, as TGA is an integral technique (Prime et al. 2009; Sergey Vyazovkin et al. 2011).

$$q = \frac{dT}{dt} \quad (13)$$

$$\left(\frac{d\alpha}{dt}\right) = \left(\frac{d\alpha}{dT}\right) q \quad (14)$$

$$\ln\left[\left(\frac{d\alpha}{dT}\right) q\right] = \ln[A\, f(\alpha)] - \frac{E_a}{RT} \quad (15)$$

Data obtained from integral techniques are better described by employing integral methods, such as those proposed by Osawa (Ozawa 1965) and Flynn and Wall (Flynn and Wall 1966b; 1966a). Integrating Equation 4 from zero to t (Equation 16), considering a constant heating rate (Equation 17, where T_0 is the temperature at t = 0 s), allows the integration of the same equation with respect to temperature (from T_0 to T), as shown in Equation 18.

$$g(\alpha) = A \int_0^t exp\left(\frac{-E_a}{RT}\right) dt \quad (16)$$

$$T = T_0 + qt \quad (17)$$

$$g(\alpha) = \frac{A}{q} \int_{T_0}^{T} exp\left(\frac{-E_a}{RT}\right) dT = \frac{A}{q} I(E_a, T) \quad (18)$$

The integral $I(E_a,T)$ in Equation 18 does not have an analytical solution, so the Osawa–Flynn–Wall (OFW) approximation (Doyle 1962) was proposed as the final form of the equation (Equation 19). From Equation 19, the slope of the curve log(q) vs. 1/T is equal to $-0.4567E_a/R$ for each α.

$$\log q = log\left[\frac{AE_a}{Rg(\alpha)}\right] - 2.315 - 0.4567\frac{E_a}{RT} \quad (19)$$

Akahira and Sunose (Akahira and Sunose 1971) proposed another integral method for calculating E_a in isoconversional analysis based on the method proposed by Kissinger (Kissinger 1957). Equation 20 represents the Kissinger–Akahira–Sunose (KAS) method. Similar to the OFW method, in the KAS method, the slope of $\log(q/T^2)$ vs. $1/T$ is equal to $-E_a/R$ for each α.

$$\ln\left(\frac{q}{T^2}\right) = const - \frac{E_a}{RT} \tag{20}$$

The Friedman, OFW, and KAS methods can be considered as classical methods. All of them approximate the integral $I(E_a, T)$, so there is always an error associated with the applications of these methods that are based on linear regression analysis. Among these three methods, the most accurate is the KAS method (Sergey Vyazovkin et al. 2011; Starink 2003). The use of numerical integration allows an even more accurate determination of the dependence of E_a with respect to α. This approach was used by Vyazovkin (Sergey Vyazovkin and Dollimore 1996; Sergey Vyazovkin 1997; 2001), whose methods use the minimization of the function presented in Equation 21, where the integral $I(E_a, T)$ (Equation 22) is solved numerically. The minimization of Equation 21 is repeated for each α value to obtain E_a for the specific conversion (Sergey Vyazovkin et al. 2011).

$$\Phi(E_a) = \sum_{i=1}^{n} \sum_{j \neq i}^{n} \frac{I(E_a, T_i) q_j}{I(E_a, T_j) q_i} \tag{21}$$

$$I(E_a, T) = \int_0^T \exp\left(\frac{-E_a}{RT}\right) dT \tag{22}$$

Most of the published works demonstrating an isoconversional kinetic computation of thermal analysis data use the simplest methods based on linear regressions (Equations 15, 19, and 20) owing to their simplicity compared with the more complex numerical computational analysis; for most cases, this is considered acceptable. More accurate methods should be considered when the E_a differs from average value by more than 20–30% (Sergey Vyazovkin et al. 2011).

When using an isoconversional methodology to calculate the kinetic parameters of thermal decomposition processes, E_a is directly calculated from the linear regressions of Equations 15, 19, and 20 or by minimizing Equation 21; however, the pre-exponential factor, A, cannot be calculated in a straightforward manner. A more appropriate way to determine A is to apply the compensation effect (S. Vyazovkin 2015; Sergey Vyazovkin et al. 2011). Using the data obtained from a single heating rate yields a series of E_{ai} and lnA_i values by applying Equation 23 and substituting $f(\alpha)$ for the kinetic models in Table 2 (each $f(\alpha)$ will provide one pair of E_{ai} and lnA_i). Although these pairs of kinetic parameters do not have a physical meaning, they are strongly correlated in the form of Equation 24. Once the values of a and b are known, and inserting the E_a values calculated for each α by employing one of the aforementioned isoconversional methods, it is possible to obtain the value of ln$A$ for each α. It is important to note that the determination of A using the compensation effect is only applicable to thermal decomposition processes that occur in a single step (Sergey Vyazovkin 2021). Having determined both E_a and A with acceptable accuracy, it is possible to determine $f(\alpha)$, which describes the experimental data (Sbirrazzuoli 2013), and to even obtain insights into the reaction mechanisms (Sbirrazzuoli 2020).

$$\frac{d\alpha}{dT}q = Aexp\left(\frac{-E_a}{RT}\right)f(\alpha) \tag{23}$$

$$\ln A_i = aE_{ai} + b \tag{24}$$

This section of the chapter provides some guidelines on how to obtain the kinetic parameters that describe thermally stimulated processes, such as thermal decomposition, using TGA. It is beyond the scope of this chapter to describe these procedures in more detail, so we strongly encourage readers to refer to other works that focus on the kinetic computation of thermal analysis data, especially the book written by S. Vyazovkin (S. Vyazovkin 2015) and the ICTAC Kinetic Committee's recommendations on how to perform these calculations (Sergey Vyazovkin et al. 2011; 2014; 2020).

Kinetic Analysis of the Thermal Decomposition of Dicationic Ionic Liquids

An analysis of the long-term thermal stability of dicationic imidazolium-based ILs by determining the kinetic parameters (E_a and A) has been performed for only a small number of ILs using both isothermal and isoconversional methodologies. For example, Frizzo et al. (Frizzo et al. 2018) isoconversionally analyzed the thermal decomposition of DILs $[C_8(MIM)_2][X]_2$ (X = Br^-, Cl^-, NTf_2^-, BF_4^-, NO_3^-, and SCN^-) using the OFW and Friedman methods. For both methods, the results indicated three different profiles for the dependence of E_a on α depending on the anion: constant over the entire range of α (X = Br^-), decreasing with increasing α (X = SCN^-), or increasing as α increased from 0.1 to 0.9 (X = Cl^-, NTf_2^-, BF_4^-, and NO_3^-). The authors attributed the variation in E_a as the reaction progressed to the breaking of the chemical bonds in the anions, resulting in competitive reactions that contributed to the overall E_a of the process, such as the formation of SO_2 in the case of NTf_2^-.

To analyze the influence of the spacer chain on the thermal stability of DILs $[C_n(MIM)_2][Br]_2$ (n = 10, 12, and 14), Bender *et al.* (Bender et al. 2019) used the OFW and Friedman isoconversional methods to determine the kinetic parameters of thermal decomposition. For both methods, E_a varied over the α range, especially in the case of $[C_{10}(MIM)_2][Br]_2$ and $[C_{14}(MIM)_2][Br]_2$. The results indicated that $[C_{12}(MIM)_2][Br]_2$ was more thermally stable than the other two ILs, indicating no direct relationship between the thermal stability and spacer-chain length of these ILs. In this study (Bender et al. 2019) and another by Frizzo et al. (Bender et al. 2019; Frizzo et al. 2018), the pre-exponential factor, *A* was determined from linear regressions of the equations obtained from the respective isoconversional methods. This is a limited approach because it is not possible to directly calculate a single value of *A* directly using Equations 15 and 19 (S. Vyazovkin 2015).

Using carboxylate anions, Vieira et al. (Vieira, Villetti, and Frizzo 2021) analyzed the kinetics of the thermal decomposition of the DILs $[C_4(MIM)_2][C_nCOO]_2$ (n = 3 and 6) and $[C_{10}(MIM)_2][C_3COO]_2$ by applying both isothermal and isoconversional methodologies. The curves of α vs. *t* in the isothermal analysis exhibited a decelerating profile and were fitted accordingly. The best results were obtained using three kinetic models: F0 ($f(\alpha) = 1$), D1 ($f(\alpha) = 1/2\alpha^{-1}$), and R2 ($f(\alpha) = 2(1-\alpha)^{1/2}$). To refine the results, the theoretical considerations of these three kinetic models were factored in, and the R2 model was considered inadequate. The best correlation coefficient between the kinetic models F0 and D1 was considered as the criterion for choosing the reported E_a. The order of thermal stability established from isothermal analysis was $[C_4(MIM)_2][C_6COO]_2$ (138 kJ mol^{-1}, kinetic model F0) > $[C_{10}(MIM)_2][C_3COO]_2$ (127 kJ mol^{-1}, kinetic model D1) > $[C_4(MIM)_2][C_3COO]_2$ (122 kJ mol^{-1}, kinetic model F0). The kinetic model D1 that best fitted the experimental data for $[C_{10}(MIM)_2][C_3COO]_2$ suggests that the thermal decomposition of this IL is regulated by the movement of the reactants, while for $[C_4(MIM)_2][C_3COO]_2$

and $[C_4(MIM)_2][C_6COO]_2$, F0 indicated that the mobility of the reagents did not affect the decomposition reaction.

Vieira et al. (Vieira, Villetti, and Frizzo 2021) performed an isoconversional analysis by applying the Friedman, OFW, and KAS methods. For $[C_4(MIM)_2][C_6COO]_2$, two well-defined decomposition stages were observed and treated separately. The results obtained from the three methods indicated that except for the first step of the decomposition of $[C_4(MIM)_2][C_6COO]_2$, E_a was dependent on α. In general, the E_a values for the ILs containing the C_3COO^- ion decreased with increasing α, except for the second step of $[C_4(MIM)_2][C_6COO]_2$, in which E_a increased with increasing α. For $[C_{10}(MIM)_2][C_3COO]_2$, two average E_a values were observed depending on the α region, indicating that this IL decomposed via a two-stage reaction. Regardless of the isoconversional method used, $[C_4(MIM)_2][C_3COO]_2$ exhibited the lowest E_a over the entire α range, indicating its lower thermal stability. Considering the first stage of thermal decomposition, $[C_4(MIM)_2][C_6COO]_2$ (first decomposition reaction) and $[C_{10}(MIM)_2][C_3COO]_2$ had very similar E_a values. The order of thermal stability obtained using isoconversional analysis was the same as in the isothermal methodology, i.e., $[C_4(MIM)_2][C_6COO]_2 > [C_{10}(MIM)_2][C_3COO]_2 > [C_4(MIM)_2][C_3COO]_2$, but was determined with higher accuracy.

In the same work, *A* was determined by applying the compensation effect for the first decomposition step of $[C_4(MIM)_2][C_6COO]_2$ (the only case of invariable E_a). However, as highlighted by Vyazovkin (Sergey Vyazovkin 2021), the use of the E_a values obtained using the OFW method as input in the compensation effect equation was not ideal owing to the low accuracy of this method. Having determined the E_a and *A* values for the first decomposition step of $[C_4(MIM)_2][C_6COO]_2$, this work also presents the determination of $f(\alpha)$ = D3 as the kinetic model that best describes the experimental data without any arbitrary assumptions.

Clarke *et al.* (Clarke et al. 2020) performed an isothermal kinetic analysis of thermal decomposition for a series of ILs $[(C_nIMC_1)_2Py][A]_2$ (n = 6, 8; A = NTf_2^- and Cl^-) assuming pseudo-first-order decomposition kinetics (kinetic model F0). The results showed a direct relationship between the molar masses and thermal stabilities of the ILs with the following order of stability: $[(C_8IMC_1)_2Py][NTf_2]_2$ (192.7 kJ mol^{-1}) > $[(C_6IMC_1)_2Py][NTf_2]_2$ (188.6 kJ mol^{-1}) > $[(C_8IMC_1)_2Py][Cl]_2$ (128.7 kJ mol^{-1}). In another study, Clarke et al. (Clarke et al. 2021) performed an isoconversional analysis for similar ILs. The authors used the OFW method to calculate the E_a of $[(C_8IMC_1)_2Py][NTf_2]_2$ as 166.3 kJ mol^{-1} ($\alpha = 0.05$), which is considerably lower than that obtained in their previous work using the isothermal methodology. In the case of $[(C_8BzIM)_2Py][NTf_2]_2$, E_a was equal to 185.7 kJ mol^{-1} for the same α, indicating that the benzimidazole cation enhances the thermal stability of the IL.

Decomposition Mechanisms

For ILs based on the imidazolium ring, the greatest cause of thermal degradation is the neutralization of the ring through second-order nucleophilic substitution (S_N2). This reaction promotes the loss of the IL alkyl chain (B. Wang et al. 2017; Maton, Vos, and Stevens 2013). Nevertheless, additional and simultaneous bond breakages occur in different ILs, which can explain the observed differences in the E_a and ln$A$ values as a function of α. As the kinetic energy of the system increases with increasing temperature, the E_a of further thermal degradation reactions can be reached. Furthermore, depending on the temperature and heating rate, the decomposition mechanisms may be subjected to multiple simultaneous reactions, autocatalytic effects, and competing reaction pathways with different rates (Maton, Vos, and Stevens 2013; Heym et al. 2011). Therefore, a specific evaluation of each structure or set of structures is mandatory, as different cations, anions, and/or functional groups can promote important changes in the decomposition mechanism of ILs. In this section, research regarding the decomposition mechanism of imidazolium-based DILs will be detailed.

Patil et al. (Patil et al. 2018) investigated the thermal stability of 15 bis-dicationic ILs derived from diimidazolium, dipyrrolidinium, or diphosphonium cations and NTf_2^- anions. Two imidazolium-based DILs with an alkyl spacer and two with a polyethylene glycol (PEG) spacer chain were synthesized. The thermal degradation products were analyzed using mass spectrometry to obtain insights into the structural linkages, bonds, and atoms most susceptible to thermally induced changes in the structures.

In general, heteroatom–carbon single bonds are inclined toward thermolytic decomposition. For ILs derived from dimethylimidazole, a peak at *m/z* 207 was observed, corresponding to the loss of a dimethylimidazole group and methylene group. A possible mechanism for this transformation was suggested by the authors and is shown in Figure 3.

The loss of alkyl substituents on the imidazole ring is attributed to nucleophilic NTf_2^- anions. Known as the reverse Menshutkin reaction, this reaction occurs for ammonium and imidazolium ILs containing halide ions

(Maton, Vos, and Stevens 2013; Sawada et al. 1985; Baranyai et al. 2004; Ngo et al. 2000). The authors suspected that the low nucleophilicity of the NTf_2^- anion (compared with halides) was the reason for the higher thermal stability of these DILs because the reverse Menshutkin reaction begins to occur at higher temperatures (>350 °C). After the first thermal degradation, methylene groups are lost from the alkyl spacer chains (*m/z* 14).

For PEG-linked DILs, the fragment-loss evaluation indicates that the decomposition (at ≈300 °C) initiates with C–O bond breaking, which is different from C–N breaking for alkyl-chain-linked DILs. Figure 4 shows the postulated mechanisms for this class of DILs.

The authors concluded that the bond breaking as the temperature increased did not follow the order of bond strength. This result can be attributed to factors such as fragment stability and the influence of substituents. Additionally, it was observed that the positions and number of substituents on the alkyl spacer chain can promote changes in the stability and degradation patterns of the DILs.

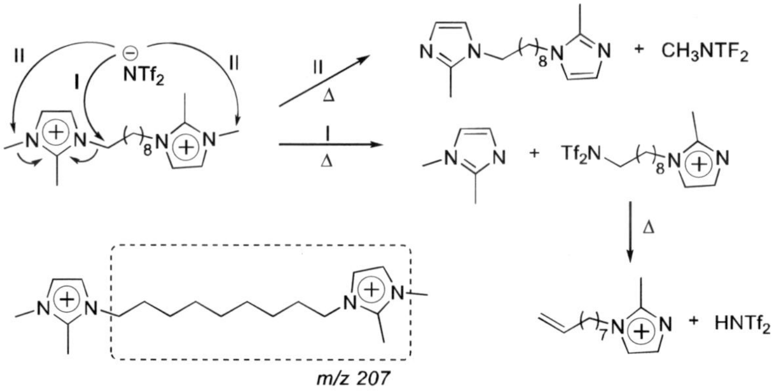

Figure 3. Decomposition mechanism for imidazolium-based DILs. Adapted from reference (Patil et al. 2018).

Vieira et al. (Vieira, Villetti, and Frizzo 2021) synthesized DILs containing carboxylate anions (bis(C_nMIM)][C_mCOO]$_2$, where n = 4 and 10, and m = 3 and 6) and determined their thermal stability and decomposition mechanisms. From the kinetic analysis of thermal decomposition using isothermal and isoconversional methodologies, decomposition mechanisms were proposed for the DILs structures. The thermal decomposition mechanism was determined by analyzing the thermal decomposition residue from isothermal analysis using different characterization tools, such as ESI-MS, ^{1}H-NMR, and a TGA instrument equipped with an FTIR detector. For the non-isothermal analysis, the gas flow from the TGA furnace was analyzed using the coupled FTIR detector. The results showed that the decomposition process involved more than one reaction, and it was found that the increase in the cation spacer chain length tended to decrease the stability.

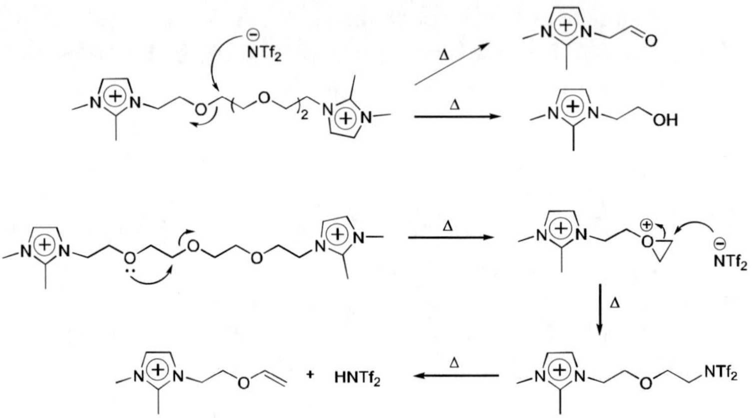

Figure 4. Decomposition mechanism for imidazolium-based PEG-linked DILs. Adapted from reference (Patil et al. 2018).

The evaluation of the thermal decomposition of DILs, for both isothermal and non-isothermal conditions, showed a variation in E_a values, indicating a multistep decomposition mechanism for these structures. The proposed pathways for the thermal decomposition of imidazolium-based ILs with carboxylate as counterions are based on the following reactions: (I) N-heterocyclic carbene (NHC) formation, (II) nucleophilic substitution at the side chain, (III) nucleophilic substitution at the methyl group, (IV) elimination reaction at the side chain, and (V) CO_2 formation, according to the mechanism shown in Figure 5. As previously reported for imidazolium-based monocationic (Clough et al. 2013; Efimova et al. 2018; Thomas et al. 2018; S. Liu et al. 2015; Rezaeian, Izadyar, and Housaindokht 2020) and dicationic (Patil et al. 2018) ILs, dications not only decompose via nucleophilic substitution reactions at both the methyl and methylene positions (mechanisms II and III)) but also through the formation of N-heterocyclic carbenes (mechanisms I) and elimination reactions (mechanisms IV). Mechanisms V represents the main decomposition route of carboxylate anions.

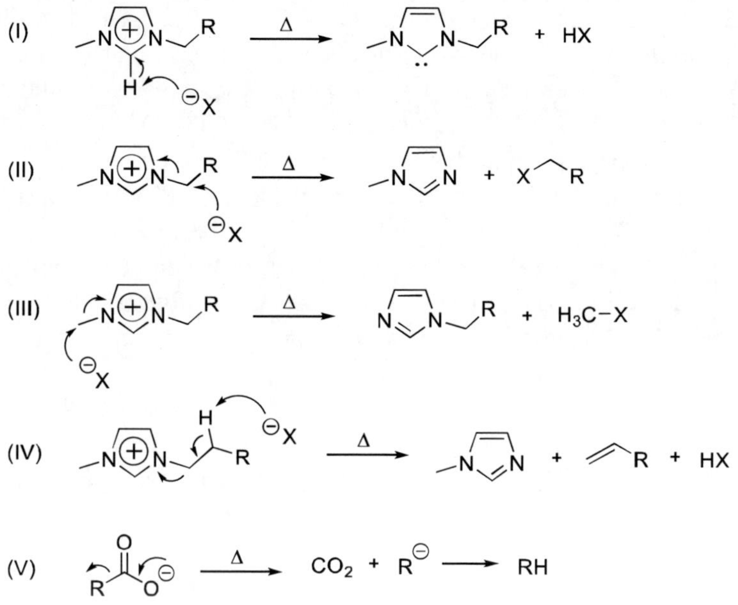

Figure 5. Thermal decomposition pathways for imidazolium-based ILs and carboxylateanions. Adapted from reference (Vieira, Villetti, and Frizzo 2021).

Clarke et al. (Clarke et al. 2020) thermally characterized a new thermally stable dicationic IL with pyridine functional groups, $[(C_8ImC_1)_2Py][A]_2$, in which A = Cl^-, PF_6^-, BF_4^-, OTf^-, NTf_2^-. The processes that led to their decomposition were evaluated and compared with those of nonfunctional geminal dicationic ILs ($[(C_nImC_1)_2][NTf]_2$; n = 3, 6, and 12). The results showed that $[(C_8ImC_1)_2Py][A]_2$ with noncoordinating anions such as NTf_2^- possessed thermal stabilities comparable to those of nonfunctional geminal dicationic ILs, with the added advantage of a functional pyridine moiety. Simultaneous Thermal Analysis (STA) was performed for $[(C_8ImC_1)_2Py][A]_2$ and $[(C_8ImC_1)_2Py][A]_2$ to obtain mechanistic information. It was found that the IL decomposed endothermically, except for $[(C_8ImC_1)_2Py][OTf]_2$, which underwent exothermic decomposition. Interestingly, the T_d peak for $[(C_nImC_1)_2][NTf]_2$ depended on the alkyl spacer length. The endothermic decomposition behavior observed for most ILs in this study was opposite to the exothermic decomposition behavior of monocationic imidazolium ILs with NTf_2^- anions (Ngo et al. 2000). The authors suggested that long-chain (n $\geq$ 3) geminal dicationic ILs tended to produce more stable decomposition products, reflecting their endothermic decomposition.

Recently, Clarke et al. (Clarke et al. 2021) performed a deeper investigation into the decomposition mechanisms of dicationic pyridine ILs. This type of IL is thermally robust, making it a promising candidate for applications involving high temperatures. The STA data suggest that multiple decomposition mechanisms occurred in parallel for these structures. Furthermore, all salts except $[(C_8Im)_2Py][OTf]_2$ endothermically decomposed. The STA signals for $[(C_8Im)_2Py][NTf_2]_2$ and $[(C_8BzIm)_2Py][NTf_2]_2$ indicated that the presence of benzimidazolium groups influenced the decomposition mechanism. For $[(C_8Im)_2Py][NTf_2]_2$, TGA–MS showed that CO_2 release occurred before any alkyl chain loss or anion degradation. This observation allowed the authors to suggest that $[(C_8Im)_2Py][NTf_2]_2$ starts to decompose by nucleophilic attack of the pyridine functional group on the NTf_2^- anion (oxygen source). On the other hand, $[(C_8BzIm)_2Py][NTf_2]_2$ thermal decomposition simultaneously produces CO_2 and alkyl chains, which suggests that the benzimidazolium salt is decomposed by alkyl chain loss (e.g., Hoffman elimination). The authors concluded that the dication was the limiting factor for IL stability because the pyridine–imidazolium C–N bond is susceptible to thermolysis owing to the electron-withdrawing effect of the imidazolium rings.

Using thermogravimetric data, Frizzo et al. (Frizzo et al. 2018) determined the thermokinetic and thermodynamic parameters of activation for a series of DILs abbreviated as $[BisOct(MIM)_2][2X]$ (where X = Cl^-, Br^-, NO_3^-, SCN^-, BF_4^-, and NTf_2^-). They found that the anions influenced the decomposition mechanism of these IL structures. For $[BisOct(MIM)_2][2Br]$, E_a and *lnA* (obtained using the OWF and Friedman methodologies) remained constant for all values of α evaluated, indicating a single-step decomposition reaction. However, for $[BisOct(MIM)_2][2X]$ (where X = NTf_2^-, BF_4^-, and NO_3^-) and $[BisOct(MIM)_2][2SCN]$, E_a and ln*A* increased and decreased, respectively. In these cases, the decomposition process occurs in multiple steps involving competitive reaction pathways with different reaction rates, which can be dependent on the temperature and heating rate. The single-step decomposition of $[BisOct(MIM)_2][2Br]$ was related to the S_N2 dealkylation reaction via nucleophilic attack of the halide in the DIL, while the multiple reactions in $[BisOct(MIM)_2][2X]$ (where X = NTf_2^-, BF_4^-, NO_3^-, and SCN^-) were associated with additional bond breakages in the IL anions. Density functional theory (DFT) calculations have shown that NTf_2^- decomposition in a monocationic IL occurs via an exothermic release of sulfur dioxide (SO_2) (Kroon et al. 2007). The authors suggest that the same exothermic SO_2 release, occurs for $[BisOct(MIM)_2][2NTf_2]$. The activation parameters governing the thermal decomposition of ILs, namely entropy ($\Delta S^{\ddagger}$), enthalpy ($\Delta H^{\ddagger}$), and Gibbs free energy ($\Delta G^{\ddagger}$), were calculated at 298.15 K using E_a and ln$A$ values corresponding to α values of 0.1 and 0.5, respectively (i.e., 10% and 50% activity loss, respectively). The positive values of the $\Delta G^{\ddagger}$ and $\Delta H^{\ddagger}$ indicate an endergonic and endothermic process, respectively, for the decomposition of these DILs, in which the decomposition reactions are non-spontaneous and dependent on heating. From these data, the following stability order was proposed for the [BisOct(MIM)$_2$][2X]: NO_3^- > BF_4^- > Br^- > SCN^- > NTf_2. This order was applicable to all α values evaluated. The low $\Delta S^{\ddagger}$ indicates the low reactivity of these DILs, in which the time required to attain the decomposition-activated complex is long (Shamsipur et al. 2013). To the best of our knowledge, to date this was the first investigation involving DILs mechanism that addressed activation parameters to obtain information regarding ILs thermal stability.

Bender et al. (Bender et al. 2019) used thermokinetics and activation parameters to determine the decomposition mechanisms of dicationic ([BisAlk(MIM)$_2$][2Br]) and monocationic ([Alk(MIM)][Br]) ILs, where Alk = Dec, DoDec, and TetDec. Their main objective was to evaluate the influence of an additional cationic head and alkyl chain length on the processes. From the TGA data, two different methodologies were applied (OWF and Friedman) to non-isothermal TGA data at different heating rates. The kinetic parameters (E_a and ln$A$) of mono- and dicationic ILs derived from both methods presented changes as a function of the mass fraction of conversion (α). This indicates a multistep decomposition mechanism with different E_a values. Additionally, mono- and dicationic ILs showed distinct trends (ascendent or descendent curves) for E_a and ln$A$ as a function of α, which was attributed to two main reasons: (I) different decomposition reactions occurring in the mono- and dicationic IL alkyl chains, and (II) decomposition of an extra imidazolium ring in the dicationic ILs. From the E_a profile as a function of α, the stability order was determined to be [Dec(MIM)][Br] > [TetDec(MIM)][Br] > [DoDec(MIM)][Br] for mono-and [BisDoDec(MIM)$_2$][2Br] > BisTetDec(MIM)$_2$][2Br] > [BisDec (MIM)$_2$][2Br] for dicationic ILs. The activation parameters evaluation (at 298.15 K) confirmed that the alkyl chain length influenced the decomposition of the mono- and dicationic ILs, although not linearly. In general, the dicationic ILs were found to be more stable than the monocationic ILs, with the exception of [Dec(MIM)][Br].

Sidelnikov et al. (Shashkov and Sidelnikov 2012) used mass spectrometry (MS) data to evaluate single-cation ILs and DILs as stationary liquid phases in GC–MS at high temperatures. Fragmentation mechanisms have also been proposed. The MS spectrum of dicationic [BisOct(MMIM)$_2$][2NTf$_2$] showed a set of low-intensity peaks related to the low extent of degradation of this DIL phase at 300 °C. Similarly to the monocationic IL analogs, [BisOct(MMIM)$_2$][2NTf$_2$] gave rise to peaks at *m/z* 64 and 69 (corresponding to anionic fragments) some peak corresponding to dicationic fragments, where the most appeared at *m/z* 124. This latter peak does not change at elevated temperatures and is related to the lack of appreciable volatility and degradation of the IL up 300 °C. This result is in agreement with that of a previous investigation conducted by Anderson et al. (Anderson et al. 2005). The dicationic imidazole-based ILs showed prominent volatility only at temperatures above 350 °C.

Conclusion

Structural modifications in spacer length or side chain of DILs, including the presence of functional groups, are very promising to design new structures with appropriated thermal stabilities to fit desired applications. It was showed that the mostly used parameter to estimate thermal stability of ILs, abbreviated as T_{onset}, only represents a short-term thermal stability. Even though this data can be compared and used to establish tendencies, real applications are directly dependent on the long-term thermal stability, obtained by application of appropriate kinetic models in decomposition data. Furthermore, a lack of standardization regarding parameters used to evaluate short-term thermal stabilities in the literature, can contribute to erroneous assumptions and non-applicable comparisons for analogous structures.

Using kinetic analyses to determine of the long-term thermal stability of DILs is more informative and accurate than using other approaches. Kinetic analyses fall into two different categories. In isothermal kinetic analysis, samples are maintained at a certain temperature for a predetermined period, and the kinetic parameters are determined by choosing a theoretical kinetic model that describes the experimental data. The use of this kinetic model is a source of error in the methodology. In addition, a single E_a value is obtained for the entire decomposition process. To avoid choosing a kinetic model, the isoconversional method is required, which is the recommended method. The use of different heating rates allows the determination of the kinetic parameters with higher accuracy, and several values of E_a can be calculated as the reaction progresses, which is useful for the analysis of the thermal degradation mechanism because it indicates if there is more than one decomposition step contributing to the overall E_a.

As reported, changes in the imidazolium cation or anion structure can promote sudden changes in the ILs properties, meaning significantly differences in the E_a and ln$A$, kinetics parameters of thermal decomposition reactions, as a function of α. In addition, to elucidate the mechanism of decomposition reactions, this understanding is pivotal to not attribute misappropriate stabilities for ILs, which can be recurrent when only short-term thermal stability parameters are considered. The use of kinetic analysis to determine the long-term thermal stability of DILs is rare in the literature, as emphasized in this chapter by the small number of studies reporting this type of data. This gap in this emerging field of research should be taken as an incentive for IL research groups to conduct kinetic analyses and expand our knowledge of the thermal properties of these compounds.

Disclaimer

None.

References

Akahira, T., and T. Sunose. 1971. "Joint Convention of Four Electrical Institutes." Science Technology, *Chiba Inst. Tecn.* 16: 22–31.

Anderson, Jared L., Rongfang Ding, Arkady Ellern, and Daniel W. Armstrong. 2005. "Structure and Properties of High Stability Geminal Dicationic Ionic Liquids." *Journal of the American Chemical Society* 127 (2): 593–604. https://doi.org/10.1021/ja046521u.

Baranyai, Krisztian J., Glen B. Deacon, Douglas R. MacFarlane, Jennifer M. Pringle, and Janet L. Scott. 2004. "Thermal Degradation of Ionic Liquids at Elevated Temperatures." *Australian Journal of Chemistry* 57 (2): 145. https://doi.org/10.1071/CH03221.

Beck, Thaíssa S., Mara de Mattos, Carlos R. T. Jortieke, Jean C. B. Vieira, Camila M. Verdi, Roberto C. V. Santos, Michele R. Sagrillo, Aline Rossato, Larissa da Silva Silveira, and Clarissa P. Frizzo. 2022. "Structural Effects of Amino Acid-Based Ionic Liquids on Thermophysical Properties, and Antibacterial and Cytotoxic Activity." *Journal of Molecular Liquids* 364 (October): 120054. https://doi.org/10.1016/j.molliq.2022.120054.

Bender, Caroline, Bruna Kuhn, Carla Farias, Francieli Ziembowicz, Thaíssa Beck, and Clarissa Frizzo. 2019. "Thermal Stability and Kinetic of Decomposition of Mono- and

Dicationic Imidazolium-Based Ionic Liquids." *Journal of the Brazilian Chemical Society* 30 (10): 2199–2209. https://doi.org/10.21577/0103-5053.20190114.

Boumediene, Mostefa, Boumediene Haddad, Annalisa Paolone, Mohammed Amin Assenine, Didier Villemin, Mustapha Rahmouni, and Serge Bresson. 2020. "Synthesis, Conformational Studies, Vibrational Spectra and Thermal Properties, of New 1,4-(Phenylenebis(Methylene) Bis(Methyl-Imidazolium) Ionic Liquids." *Journal of Molecular Structure* 1220 (November). https://doi.org/10.1016/j.molstruc.2020.128731.

Boumediene, Mostefa, Boumediene Haddad, Annalisa Paolone, Mokhtar Drai, Didier Villemin, Mustapha Rahmouni, Serge Bresson, and Ouissam Abbas. 2019. "Synthesis, Thermal Stability, Vibrational Spectra and Conformational Studies of Novel Dicationic Meta-Xylyl Linked Bis-1-Methylimidazolium Ionic Liquids." *Journal of Molecular Structure* 1186 (June): 68–79. https://doi.org/10.1016/j.molstruc.2019.03.019.

Cao, Yuanyuan, and Tiancheng Mu. 2014. "Comprehensive Investigation on the Thermal Stability of 66 Ionic Liquids by Thermogravimetric Analysis." *Industrial and Engineering Chemistry Research* 53 (20): 8651–64. https://doi.org/10.1021/ie5009597.

Clarke, Coby J., Liem Bui-Le, Jason P. Hallett, and Peter Licence. 2020. "Thermally-Stable Imidazolium Dicationic Ionic Liquids with Pyridine Functional Groups." *ACS Sustainable Chemistry & Engineering* 8 (23): 8762–72. https://doi.org/10.1021/acssuschemeng.0c02473.

Clarke, Coby J., Patrick J. Morgan, Jason P. Hallett, and Peter Licence. 2021. "Linking the Thermal and Electronic Properties of Functional Dicationic Salts with Their Molecular Structures." *ACS Sustainable Chemistry & Engineering* 9 (18): 6224–34. https://doi.org/10.1021/acssuschemeng.0c08564.

Claros, Martha, Teófilo A. Graber, Iván Brito, Joselyn Albanez, and José A. Gavín. 2010. "SyntheSiS and Thermal PropertieS of Two New Dicationic Ionic LiquidS." *J. Chil. Chem. Soc.* Vol. 55.

Clough, Matthew T., Karolin Geyer, Patricia A. Hunt, Jürgen Mertes, and Tom Welton. 2013. "Thermal Decomposition of Carboxylate Ionic Liquids: Trends and Mechanisms." *Physical Chemistry Chemical Physics* 15 (47): 20480–95. https://doi.org/10.1039/c3cp53648c.

D'Anna, Francesca, H. Q. Nimal Gunaratne, Giuseppe Lazzara, Renato Noto, Carla Rizzo, and Kenneth R. Seddon. 2013. "Solution and Thermal Behaviour of Novel Dicationic Imidazolium Ionic Liquids." *Organic and Biomolecular Chemistry* 11 (35): 5836–46. https://doi.org/10.1039/c3ob40807h.

Doyle, C. D. 1962. "Estimating Isothermal Life from Thermogravimetric Data." *Journal of Applied Polymer Science* 6 (24): 639–42. https://doi.org/10.1002/app.1962.070062406.

Efimova, Anastasia, Janos Varga, Georg Matuschek, Mohammad R. Saraji-Bozorgzad, Thomas Denner, Ralf Zimmermann, and Peer Schmidt. 2018. "Thermal Resilience of Imidazolium-Based Ionic Liquids - Studies on Short- and Long-Term Thermal Stability and Decomposition Mechanism of 1-Alkyl-3-Methylimidazolium Halides by Thermal Analysis and Single-Photon Ionization Time-of-Flight Mass Spectrometry."

Journal of Physical Chemistry B 122 (37): 8738–49. https://doi.org/10.1021/acs.jpcb.8b06416.

Fareghi-Alamdari, Reza, Razieh Hatefipour, Mehdi Rakhshi, and Negar Zekri. 2016. "Novel Diol Functionalized Dicationic Ionic Liquids: Synthesis, Characterization and DFT Calculations on H-Bonding Influence on Thermophysical Properties." *RSC Advances* 6 (82): 78636–47. https://doi.org/10.1039/c6ra17188e.

Flynn, Joseph H., and Leo A. Wall. 1966a. "A Quick, Direct Method for the Determination of Activation Energy from Thermogravimetric Data." *Polymer Letters* 4: 323–28.

———. 1966b. "General Treatment of the Thermogravimetry of Polymers." JOURNAL OF RESEARCH of the National Bureau of Standards - A. *Physics and Chemistry* 70A (6): 487–523. https://doi.org/10.6028/jres.070A.043.

Friedman, Henry L. 1964. "Kinetics of Thermal Degradation of Char-Forming Plastics from Thermogravimetry. Application to a Phenolic Plastic." *Journal of Polymer Science Part C* 6 (1): 183–95. https://doi.org/10.1002/polc.5070060121.

Frizzo, Clarissa P., Caroline R. Bender, Paulo R. S. Salbego, Carla A. A. Farias, Thayanara C. da Silva, Sílvio T. Stefanello, Tássia L. da Silveira, Félix A. A. Soares, Marcos A. Villetti, and Marcos A. P. Martins. 2018. "Impact of Anions on the Partition Constant, Self-Diffusion, Thermal Stability, and Toxicity of Dicationic Ionic Liquids." *ACS Omega 3* (1): 734–43. https://doi.org/10.1021/acsomega.7b01873.

Frizzo, Clarissa P., Aniele Z. Tier, Caroline R. Bender, Izabelle M. Gindri, Marcos A. Villetti, Nilo Zanatta, Helio G. Bonacorso, and Marcos A. P. Martins. 2015. "Structural and Physical Aspects of Ionic Liquid Aggregates in Solution." In *Ionic Liquids - Current State of the Art*, edited by Scott Handy, 1st ed., 1–40. Rijeka: InTech. https://doi.org/10.5772/59287.

Heym, Florian, Bastian J. M. Etzold, Christoph Kern, and Andreas Jess. 2011. "Analysis of Evaporation and Thermal Decomposition of Ionic Liquids by Thermogravimetrical Analysis at Ambient Pressure and High Vacuum." *Green Chemistry* 13 (6): 1453–66. https://doi.org/10.1039/C0GC00876A.

Khan, Amir Sada, Zakaria Man, Annie Arvina, Mohammad Azmi Bustam, Asma Nasrullah, Zahoor Ullah, Ariyanti Sarwono, and Nawshad Muhammad. 2017. "Dicationic Imidazolium Based Ionic Liquids: Synthesis and Properties." *Journal of Molecular Liquids* 227 (February): 98–105. https://doi.org/10.1016/j.molliq.2016.11.131.

Kissinger, Homer E. 1957. "Reaction Kinetics in Differential Thermal Analysis." *Analytical Chemistry* 29 (11): 1702–6. https://doi.org/10.1021/ac60131a045.

Kosmulski, Marek, Jan Gustafsson, and Jarl B. Rosenholm. 2004. "Thermal Stability of Low Temperature Ionic Liquids Revisited." *Thermochimica Acta* 412 (1–2): 47–53. https://doi.org/10.1016/j.tca.2003.08.022.

Kroon, Maaike C., Wim Buijs, Cor J. Peters, and Geert Jan Witkamp. 2007. "Quantum Chemical Aided Prediction of the Thermal Decomposition Mechanisms and Temperatures of Ionic Liquids." *Thermochimica Acta* 465 (1–2): 40–47. https://doi.org/10.1016/j.tca.2007.09.003.

Kuhn, Bruna L., Bárbara F. Osmari, Thaíse M. Heinen, Helio G. Bonacorso, Nilo Zanatta, Steven O. Nielsen, Dineli T. S. Ranathunga, Marcos A. Villetti, and Clarissa P. Frizzo.

2020. "Dicationic Imidazolium-Based Dicarboxylate Ionic Liquids: Thermophysical Properties and Solubility." *Journal of Molecular Liquids* 308 (June). https://doi.org/10.1016/j.molliq.2020.112983.

Lever, Trevor, Peter Haines, Jean Rouquerol, Edward L. Charsley, Paul van Eckeren, and Donald J. Burlett. 2014. "ICTAC Nomenclature of Thermal Analysis (IUPAC Recommendations 2014)." *Pure and Applied Chemistry* 86 (4): 545–53. https://doi.org/10.1515/pac-2012-0609.

Li, Xiaoju, Duncan W. Bruce, and Jean'Ne M. Shreeve. 2009. "Dicationic Imidazolium-Based Ionic Liquids and Ionic Liquid Crystals with Variously Positioned Fluoro Substituents." *Journal of Materials Chemistry* 19 (43): 8232–38. https://doi.org/10.1039/b912873e.

Liu, Jianlian, Wenge Yang, Zhuo Li, Fei Ren, and Hong Hao. 2020. "Experimental Investigation of Thermo-Physical Properties of Geminal Dicationic Ionic Compounds for Latent Thermal Energy Storage." *Journal of Molecular Liquids* 307 (June). https://doi.org/10.1016/j.molliq.2020.112994.

Liu, Shuangyue, Yu Chen, Yang Shi, Haitao Sun, Zhengyu Zhou, and Tiancheng Mu. 2015. "Investigations on the Thermal Stability and Decomposition Mechanism of an Amine-Functionalized Ionic Liquid by TGA, NMR, TG-MS Experiments and DFT Calculations." *Journal of Molecular Liquids* 206: 95–102. https://doi.org/10.1016/j.molliq.2015.02.022.

Maton, Cedric, Nils D. Vos, and Christian v Stevens. 2013. "Ionic Liquid Thermal Stabilities: Decomposition Mechanisms and Analysis Tools." *Chemical Society Reviews* 42 (13): 5963–77. https://doi.org/10.1039/c3cs60071h.

Nessim, M. I., M. T. Zaky, and M. A. Deyab. 2018. "Three New Gemini Ionic Liquids: Synthesis, Characterizations and Anticorrosion Applications." *Journal of Molecular Liquids* 266 (September): 703–10. https://doi.org/10.1016/j.molliq.2018.07.001.

Ngo, Helen L, Karen LeCompte, Liesl Hargens, and Alan B McEwen. 2000. "Thermal Properties of Imidazolium Ionic Liquids." *Thermochimica Acta* 357–358 (August): 97–102. https://doi.org/10.1016/S0040-6031(00)00373-7.

Ozawa, Takeo. 1965. "A New Method of Analyzing Thermogravimetric Data." *Bull. Chem. Soc. Jpn.* 38 (1): 1881–86.

Patil, Rahul A., Mohsen Talebi, Alain Berthod, and Daniel W. Armstrong. 2018. "Dicationic Ionic Liquid Thermal Decomposition Pathways." *Analytical and Bioanalytical Chemistry* 410 (19): 4645–55. https://doi.org/10.1007/s00216-018-0878-0.

Patil, Rahul A., Mohsen Talebi, Chengdong Xu, Sumit S. Bhawal, and Daniel W. Armstrong. 2016. "Synthesis of Thermally Stable Geminal Dicationic Ionic Liquids and Related Ionic Compounds: An Examination of Physicochemical Properties by Structural Modification." *Chemistry of Materials* 28 (12): 4315–23. https://doi.org/10.1021/acs.chemmater.6b01247.

Plechkova, Natalia v, and Kenneth R Seddon. 2008. "Applications of Ionic Liquids in the Chemical Industry." *Chemical Society Reviews* 37 (1): 123–50. https://doi.org/10.1039/b006677j.

Prime, R. B., H. E. Bair, S. Vyazovkin, P. K. Gallagher, and A. Riga. 2009. "Thermogravimetric Analysis (TGA)." In *Thermal Analysis of Polymers:*

Fundamentals and Applications, edited by J. D. Menczel and R. B. Prime, 241–317. John Wiley & Sons, Inc. https://doi.org/10.1007/978-3-030-11599-9_7.

Rezaeian, Mojtaba, Mohammad Izadyar, and Mohammad Reza Housaindokht. 2020. "Thermal Decomposition Mechanisms of Some Amino Acid Ionic Liquids: Molecular Approach." *Journal of Molecular Liquids* 302 (March). https://doi.org/10.1016/j.molliq.2020.112505.

Sawada, Masami, Yoshio Takai, Chang Chong, Terukiyo Hanafusa, Soichi Misumi, and Yuho Tsuno. 1985. "PYRIDINIUM ION REACTIVITIES: SUBSTITUENT EFFECT QN THE REmRSE MENSCHUTKIN REACTION OF l-METHYLPYRIDINIUM CATIONS WITH IODIDE ANION." *Tetrahedron Letters*. Vol. 26.

Sbirrazzuoli, Nicolas. 2013. "Determination of Pre-Exponential Factors and of the Mathematical Functions f(α) or G(α) That Describe the Reaction Mechanism in a Model-Free Way." *Thermochimica Acta* 564: 59–69. https://doi.org/10.1016/j.tca.2013.04.015.

———. 2020. "Determination of Pre-Exponential Factor and Reaction Mechanism in a Model-Free Way." *Thermochimica Acta* 691 (June). https://doi.org/10.1016/j.tca.2020.178707.

Shamsipur, Mojtaba, Seied Mahdi Pourmortazavi, Ali Akbar Miran Beigi, Rouhollah Heydari, and Mina Khatibi. 2013. "Thermal Stability and Decomposition Kinetic Studies of Acyclovir and Zidovudine Drug Compounds." *AAPS PharmSciTech* 14 (1): 287–93. https://doi.org/10.1208/s12249-012-9916-y.

Shashkov, M. v., and V. N. Sidelnikov. 2012. "Mass Spectral Evaluation of Column Bleeding for Imidazolium-Based Ionic Liquids as GC Liquid Phases." *Analytical and Bioanalytical Chemistry* 403 (9): 2673–82. https://doi.org/10.1007/s00216-012-6020-9.

Shirota, Hideaki, Toshihiko Mandai, Hiroki Fukazawa, and Tatsuya Kato. 2011. "Comparison between Dicationic and Monocationic Ionic Liquids: Liquid Density, Thermal Properties, Surface Tension, and Shear Viscosity." *Journal of Chemical and Engineering Data* 56 (5): 2453–59. https://doi.org/10.1021/je2000183.

Siedlecka, Ewa Maria, Magorzata Czerwicka, Stefan Stolte, and Piotr Stepnowski. 2011. "Stability of Ionic Liquids in Application Conditions." *Current Organic Chemistry*. Vol. 15.

Slopiecka, Katarzyna, Pietro Bartocci, and Francesco Fantozzi. 2012. "Thermogravimetric Analysis and Kinetic Study of Poplar Wood Pyrolysis." *Applied Energy* 97: 491–97. https://doi.org/10.1016/j.apenergy.2011.12.056.

Starink, M. J. 2003. "The Determination of Activation Energy from Linear Heating Rate Experiments: A Comparison of the Accuracy of Isoconversion Methods." *Thermochimica Acta* 404 (1–2): 163–76. https://doi.org/10.1016/S0040-6031(03)00144-8.

Talebi, Mohsen, Rahul A. Patil, and Daniel W. Armstrong. 2018. "Physicochemical Properties of Branched-Chain Dicationic Ionic Liquids." *Journal of Molecular Liquids* 256 (April): 247–55. https://doi.org/10.1016/j.molliq.2018.02.016.

Talebi, Mohsen, Rahul A. Patil, Leonard M. Sidisky, Alain Berthod, and Daniel W. Armstrong. 2018. "Variation of Anionic Moieties of Dicationic Ionic Liquid GC

Stationary Phases: Effect on Stability and Selectivity." *Analytica Chimica Acta* 1042 (December): 155–64. https://doi.org/10.1016/j.aca.2018.07.047.

Thomas, Eapen, Deepthi Thomas, S. Bhuvaneswari, K. P. Vijayalakshmi, and Benny K. George. 2018. "1-Hexadecyl-3-Methylimidazolium Chloride: Structure, Thermal Stability and Decomposition Mechanism." *Journal of Molecular Liquids* 249 (January): 404–11. https://doi.org/10.1016/j.molliq.2017.11.029.

Ullah, Zahoor, M. Azmi Bustam, Zakaria Man, and Amir Sada Khan. 2016. "Thermal Stability and Kinetic Study of Benzimidazolium Based Ionic Liquid." In *Procedia Engineering,* 148:215–22. Elsevier Ltd. https://doi.org/10.1016/j.proeng.2016.06.577.

Vieira, Jean C. B., Alisson V. Paz, Bruno L. Hennemann, Bruna L. Kuhn, Caroline R. Bender, Alexandre R. Meyer, Anderson B. Pagliari, Marcos A. Villetti, and Clarissa P. Frizzo. 2020. "Effect of Large Anions in Thermal Properties and Cation-Anion Interaction Strength of Dicationic Ionic Liquids." *Journal of Molecular Liquids* 298 (January). https://doi.org/10.1016/j.molliq.2019.112077.

Vieira, Jean C. B., Marcos A. Villetti, and Clarissa P. Frizzo. 2021. "Thermal Stability and Decomposition Mechanism of Dicationic Imidazolium-Based Ionic Liquids with Carboxylate Anions." *Journal of Molecular Liquids* 330 (May): 115618. https://doi.org/10.1016/j.molliq.2021.115618.

Vyazovkin, S. 2015. *Isoconversional Kinetics of Thermally Stimulated Precesses*. Springer, Cham. https://doi.org/10.1007/978-3-319-14175-6.

Vyazovkin, Sergey. 1997. "Evaluation of Activation Energy of Thermally Stimulated Solid-State Reactions under Arbitrary Variation of Temperature." *Journal of Computational Chemistry* 18 (3): 393–402. https://doi.org/10.1002/(SICI)1096-987X(199702)18:3<393::AID-JCC9>3.0.CO;2-P.

———. 2001. "Modification of the Integral Isoconversional Method to Account for Variation in the Activation Energy." *Journal of Computational Chemistry* 22 (2): 178–83. https://doi.org/10.1002/1096-987x(20010130)22:2<178::aid-jcc5>3.0.co;2-%23.

———. 2021. "Determining Preexponential Factor in Model-Free Kinetic Methods: How and Why?" *Molecules* 26 (11): 3077. https://doi.org/10.3390/molecules26113077.

Vyazovkin, Sergey, Alan K. Burnham, José M. Criado, Luis A. Pérez-Maqueda, Crisan Popescu, and Nicolas Sbirrazzuoli. 2011. "ICTAC Kinetics Committee Recommendations for Performing Kinetic Computations on Thermal Analysis Data." *Thermochimica Acta* 520 (1–2): 1–19. https://doi.org/10.1016/j.tca.2011.03.034.

Vyazovkin, Sergey, Alan K. Burnham, Loic Favergeon, Nobuyoshi Koga, Elena Moukhina, Luis A. Pérez-Maqueda, and Nicolas Sbirrazzuoli. 2020. "ICTAC Kinetics Committee Recommendations for Analysis of Multi-Step Kinetics." *Thermochimica Acta* 689 (May). https://doi.org/10.1016/j.tca.2020.178597.

Vyazovkin, Sergey, Konstantinos Chrissafis, Maria Laura Di Lorenzo, Nobuyoshi Koga, Michèle Pijolat, Bertrand Roduit, Nicolas Sbirrazzuoli, and Joan Josep Suñol. 2014. "ICTAC Kinetics Committee Recommendations for Collecting Experimental Thermal Analysis Data for Kinetic Computations." *Thermochimica Acta* 590: 1–23. https://doi.org/10.1016/j.tca.2014.05.036.

Vyazovkin, Sergey, and David Dollimore. 1996. "Linear and Nonlinear Procedures in Isoconversional Computations of the Activation Energy of Nonisothermal Reactions in Solids." *Journal of Chemical Information and Computer Sciences* 36 (1): 42–45. https://doi.org/10.1021/ci950062m.

Wang, Binshen, Li Qin, Tiancheng Mu, Zhimin Xue, and Guohua Gao. 2017. "Are Ionic Liquids Chemically Stable?" Chemical Reviews. *American Chemical Society*. https://doi.org/10.1021/acs.chemrev.6b00594.

Wang, Guowei, Xiaoqing Xu, Yu Sun, Linghua Zhuang, and Cheng Yao. 2019. "Relationship between Structure and Biodegradability of Gemini Imidazolium Surface Active Ionic Liquids." *Journal of Molecular Liquids* 278 (March): 145–55. https://doi.org/10.1016/j.molliq.2018.12.066.

Wooster, Tim J., Katarina M. Johanson, Kevin J. Fraser, Douglas R. MacFarlane, and Janet L. Scott. 2006. "Thermal Degradation of Cyano Containing Ionic Liquids." *Green Chemistry* 8 (8): 691–69. https://doi.org/10.1039/b606395k.

Xu, Chenqian, and Zhenmin Cheng. 2021. "Thermal Stability of Ionic Liquids: Current Status and Prospects for Future Development." *Processes. MDPI AG*. https://doi.org/10.3390/pr9020337.

Zaky, M. T., M. I. Nessim, and M. A. Deyab. 2019. "Synthesis of New Ionic Liquids Based on Dicationic Imidazolium and Their Anti-Corrosion Performances." *Journal of Molecular Liquids* 290 (September). https://doi.org/10.1016/j:molliq.2019.111230.

Zhang, Hang, Jingru Liu, Mingtao Li, and Bolun Yang. 2018. "Functional Groups in Geminal Imidazolium Ionic Compounds and Their Influence on Thermo-Physical Properties." *Journal of Molecular Liquids* 269 (November): 738–45. https://doi.org/10.1016/j.molliq.2018.08.037.

Zhang, Hang, Wei Xu, Jingru Liu, Mingtao Li, and Bolun Yang. 2019. "Thermophysical Properties of Dicationic Imidazolium-Based Ionic Compounds for Thermal Storage." *Journal of Molecular Liquids* 282 (May): 474–83. https://doi.org/10.1016/j.molliq.2019.03.012.

Zhuang, Ling Hua, Kai Hua Yu, Guo Wei Wang, and Cheng Yao. 2013. "Synthesis and Properties of Novel Ester-Containing Gemini Imidazolium Surfactants." *Journal of Colloid and Interface Science* 408 (1): 94–100. https://doi.org/10.1016/j.jcis.2013.07.029.

Chapter 2

Chiral Ionic Liquids as Organocatalysts for Asymmetric Organic Transformations

Geeta Devi Yadav[1,*]
Priyanka Jhajharia[2,*]
and Surendra Singh[1,†]
[1]Department of Chemistry, University of Delhi, India
[2]Department of Chemistry, Kirori Mal College, University of Delhi, India

Abstract

Chiral ionic liquids are a distinct class of organic compounds derived from the chiral compounds having ionic character and their melting points are below 100°C. Ionic liquids are a non-volatile and good choice of solvent for organic transformations and have been used as an alternative green solvent in various organic transformations. Over the last two decades, the application of ionic liquids as a solvent or chiral organocatalysts anchored on ionic liquids has been established in the field of asymmetric catalysis, and mostly imidazole-derived ionic liquids were widely used for this purpose. In this chapter, we are describing the synthesis of chiral ionic liquids and their application in asymmetric organic transformations such as asymmetric Aldol reaction, cross-aldol reaction of aldehydes with ketones, Friedel-Crafts reaction, Diels-Alder reaction, asymmetric Michael addition, asymmetric reduction of prochiral ketones, asymmetric Biginelli reaction and miscellaneous reactions. The chiral ionic liquids tagged organocatalyst are polar and not soluble in a non-polar solvent and this property makes them recoverable and reusable.

* Both authors have equal contributions.
† Corresponding Author's Email: ssingh1@chemistry.du.ac.in.

In: Imidazolium
Editor: Stephen A. Reyes
ISBN: 979-8-89113-425-6

Keywords: proline, prolinamide, prolinol, MacMillan catalyst, chiral ionic liquids, asymmetric catalysis

1. Introduction

Ionic liquids (ILs) are compounds that are entirely made up of ions and have a melting point below 100°C. The distinction between molten salts and ionic liquids is based on melting point. Generally, ionic liquids and molten salts are described as liquid compounds that have ionic-covalent crystalline structures [1, 2]. Paul Walden reported the first room-temperature IL (ethyl ammonium nitrate) in 1914 and almost after one century, ILs became a major scientific research area. The number of SCI research papers published on ILs has exponentially increased from 1996 to 2022, which indicates exceeding the annual growth rates of other popular research areas. A multidisciplinary study on ILs has been developed including chemistry, chemical engineering, environmental science, and materials science. Some important fundamental points are now different from the original concepts to understand the nature of ILs, the physicochemical properties of ILs are now recognized as ranging broadly from the oft-quoted "non-flammable, nonvolatile, air, and water stable" to those that are distinctly flammable, volatile, and unstable. Numerous combinations of cations and anions encounter the definition of ILs, leading to a diverse suite of behaviors. Regardless, ILs remain more desirable than conventional volatile solvents and/or catalysts in many chemical and physical processes, often exhibiting "green" and "designer" properties to a suitable degree. The first-generation ILs having moisture-sensitive mostly contained tetrachloroaluminate anions and Wilkes developed the air and water-stable second-generation ILs which were prepared by replacing tetrachloroaluminate anions with other anions [3]. The concept of third-generation is a task-specific IL was developed by Davis [4] in 2004 (Figure 1). Based on chemical varieties of ILs have been further classified into several categories, e.g., room-temperature ILs (RTILs) [3−8], task-specific ILs (TSILs) [9, 10], polyionic liquids (PILs) [11, 12] and membranes supported IL (SILMs) [13, 14] that include composites of metal−organic frameworks supported ILs (MOFs) [15, 16].

By the changing structural motif of cation and anion, the ILs can show various physical properties. Ionic liquids were used as solvents in various important organic transformations reported in the areas of synthetic chemistry [17], catalysis [18], biocatalysis [19], and electrochemistry [20]. The ionic liquids are polar in nature therefore the product of the reaction may be separated more easily from an IL by using nonpolar solvents.

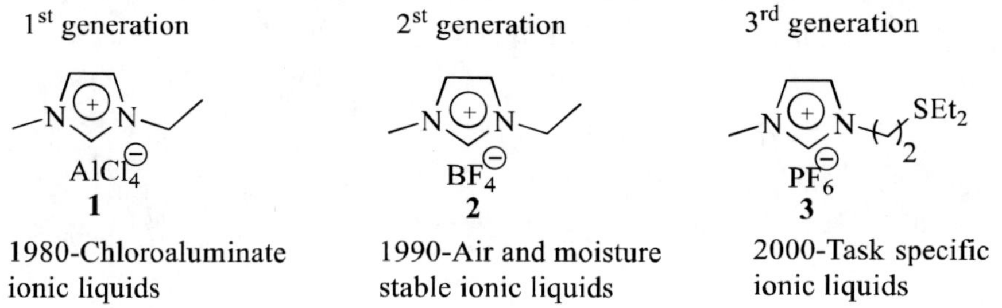

1st generation

$AlCl_4^{\ominus}$
1
1980-Chloroaluminate ionic liquids

2st generation

$BF_4^{\ominus}$
2
1990-Air and moisture stable ionic liquids

3rd generation

SEt_2
$PF_6^{\ominus}$
3
2000-Task specific ionic liquids

Figure 1. Generations of the Ionic liquid.

1.1. Chiral Ionic Liquids

The chiral ionic liquid is prepared from the chiral compounds as starting material. The chiral ionic liquid is an important branch of chemistry, it has both chiral and ionic liquid functions, which has attracted the wide attention of researchers in recent years. Hermann et al. reported N-heterocyclic carbenes of imidazole as an ionic liquid in 1996 [21]. Howarth et al. first time gave the namechiral ionic liquid in 1997 [22], synthesized chiral dialkyl imidazolium bromide, and used as the Lewis acid catalyst in the Diels-Alder reaction, affording high enantioselectivity of the product. The amino acids, borates, lactic acid, and camphorsulfonates are examples of chiral anions and the components of ionic liquids. The first chiral anion 1-butyl-3-methylimidazolium lactate was reported by Seddon et al. in 1999 [23]. The chiral anion containing CILs was synthesized by anion exchange between 1-methyl-3-butylimidazolium chloride and *S*-2-hydroxy propionate. However, Seddon's research has shown that chiral cation and anion does not provide chiral induction. The chiral ionic liquid must have an appropriate structure motif to bind to the enantiomer. Therefore, in recent years to ensure the binding site, many synthetic methods have been reported for the synthesis of chiral ionic liquids containing amino acids having various functional groups such as amino, thiol, hydroxyl, thioether, carboxyl, and amide [24-28]. The carbohydrates, alkaloids and amino acids derived CILs have been reported [29-31]. The CILs used in various applications such as asymmetric synthesis, chromatography, spectroscopy, extraction, electrochemistry, membrane separation, biotechnological areas, and liquid crystal. The environmentally renewable and sustainable raw materials like amino acids, sugars menthol, nicotine, etc. introduced the center of chirality of CILs.

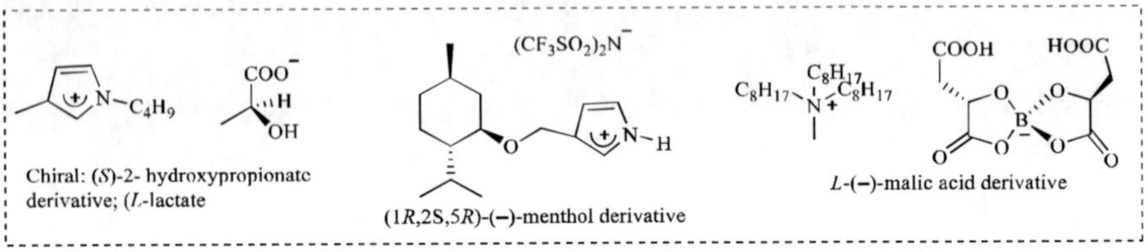

Figure 2. The general structures of chiral ionic liquids.

1.2. Chiral Ionic Liquid Catalyzed Asymmetric Aldol Reaction

The Aldol reaction is recognized as one of the most powerful carbon-carbon bond-forming reactions in modern organic synthesis. Direct asymmetric aldol reactions catalyzed by (*L*)-proline as an organocatalyst was reported by List in 2000. (*L*)-Proline as an organocatalyst was used in a substantial quantity, in some cases up to 30 mol%, and used in nonvolatile solvents like DMSO. To overcome these issues several research groups have been developed modified organocatalysts such as prolinamides which can be used in low catalytic amounts for direct asymmetric aldol reaction. To recover these organocatalysts, the catalyst was tagged with IL and can be used in low catalytic amounts for aldol reactions. The recovery and reuse of the catalyst became dynamic for reducing the cost and facilitating the separation of the product from the catalyst.

Luo et al. explored a series of chiral pyrrolidine units containing functional ionic liquid (FIL) 4 and their evaluation as reusable organocatalysts for direct asymmetric aldol reaction. FIL 4 efficiently catalyzed the aldol reaction between cyclohexanone and benzaldehyde/substituted benzaldehyde affording the corresponding aldol product with excellent yield and moderate stereoselectivities. The FIL 4 is highly reactive in the combination of water and acetic acid as an additive in the aldol reaction of ketones and aldehyde without the formation of condensation products and easily recyclable and reusable for up to six cycles with slightly reduced reactivity (Scheme 1) [32].

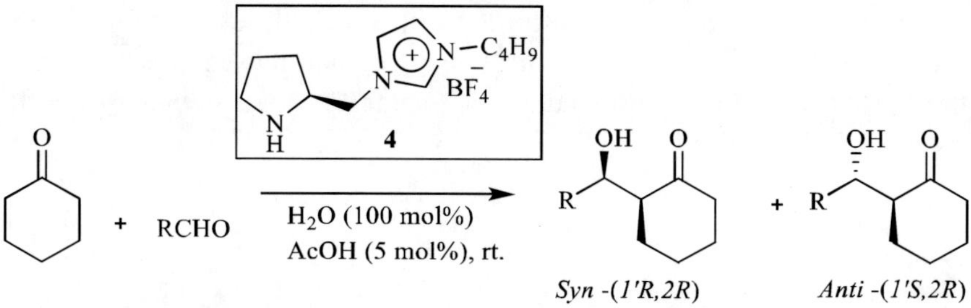

Cyclohexanone + RCHO → (catalyst **4**, H_2O (100 mol%), AcOH (5 mol%), rt.) → *Syn -(1'R,2R)* + *Anti -(1'S,2R)*

R = *o*-NO_2Ph
yield = >99%, *anti:syn* = 1.1:1
anti; ee = 32%, *Syn; ee* = 63%

R = *m*-NO_2Ph
yield = 92%, *anti:syn* = 1.1:1
anti; ee = 21%, *Syn; ee* = 46%

R = *p*-NO_2Ph
yield = 93%, *anti:syn* = 1:1
anti; ee = 10%, *Syn; ee* = 41%

R = *o*-ClPh
yield = 68%, *anti:syn* = 1.1:1
anti; ee = 20%, *Syn; ee* = 32%

R = Ph
yield = 66%, *anti:syn* = 1.3:1
anti; ee = 30%, *Syn; ee* = 46%

Scheme 1. Direct aldol reaction of cyclohexanone and 4-nitro-benzaldehyde by using FIL 4.

Zlotin and co-workers designed the hydrophilic and hydrophobic ionic liquids 5 and 6 containing prolinamide and used them as organocatalysts in direct asymmetric aldol reactions of cyclohexanone and 4-nitro-benzaldehyde in the presence of water and afforded aldol product in high yields and stereoselectivities. Hydrophobic IL-tagged organocatalyst 6 shows better activity and stereoselectivity than hydrophilic IL-tagged organocatalyst 5 in water (Scheme 2). The organocatalyst was recovered and reused for up to four cycles and the enantioselectivity and diastereoselectivity of the aldol product were retained for at least three cycles [33].

Khan has developed chiral ionic liquids tagged amino acids based on 1,2,3- trizolium 7a-c and 8 and IL 7a and 7c have additional carboxylic groups. In order to compare the effect of acidity of the carboxylic group author also synthesized the corresponding methyl ester 7b. Organocatalyst 7c has both an acylic α-amino acid and proline, while organocatalyst 8 contains long chain acyclic (*S*)-amino acid. The CILs were evaluated in a direct asymmetric aldol reaction of cyclohexanone and 4-nitrobenzaldehyde. Ester 7b gave a better result (54% *ee*) as compared to a carboxylic group containing 7a (12% *ee*) due to a non-stereoselective competing catalysis exerting by the carboxylic group. Tetrafluoroborate anion containing lysine-derived 1,2,3-triazole (8) efficiently catalyzed the aldol reaction and afforded the product in excellent yield and stereoselectivities. CIL 8 is successfully recycled and reused after extraction of the product with diethyl ether and more than 90% *ee* was obtained after the fifth run (Scheme 3) [34].

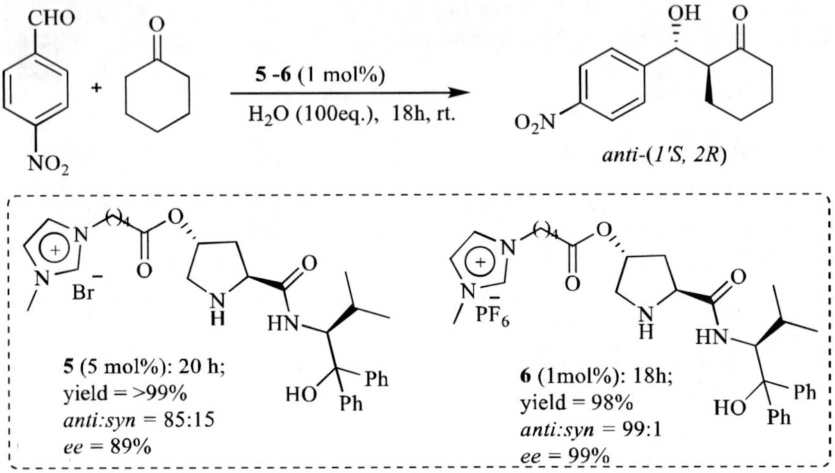

CHO
O
+
NO_2
5 -**6** (1 mol%)
H_2O (100eq.), 18h, rt.
OH O
O_2N
anti-(*1'S, 2R*)

N
N
Br
O
O
N H
O
HN
HO
Ph
Ph
5 (5 mol%): 20 h;
yield = >99%
anti:syn = 85:15
ee = 89%

N
N
PF_6
O
O
N H
O
HN
HO
Ph
Ph
6 (1mol%): 18h;
yield = 98%
anti:syn = 99:1
ee = 99%

Scheme 2. Ionic liquid modified prolinamide catalysed aldol reaction.

Zlotin and co-workers designed the first C_2-symmetric CILs 9 and 10 from prolinamide and imidazole containing hydrophilic bromide anion and hydrophobic hexafluorophosphate anion, respectively. The CILs 9 and 10 (5-10 mol%) act as an efficient catalysts for the direct asymmetric aldol reaction of cyclohexanone and 4-nitrobenzaldehyde in water and gave *anti-aldol* as the major product. However, the rate of aldol reaction and stereoselectivities rely on the nature of the anion present, and CIL 10 with hydrophobic hexafluorophosphate anion shows higher activity (conversion >99%) than CIL 9 contains hydrophilic bromide (conversion 48%) under the identical reaction condition. CIL also catalyzed the linear or cyclic ketones reacted with (hetero)- aromatic aldehydes in aqueous media to yield chiral aldol products with high diastereo- and enantioselectivities. Organocatalysts can be easily recycled and reused up to 10 times with retained reactivity and enantioselectivity of the aldol product (Scheme 4) [35].

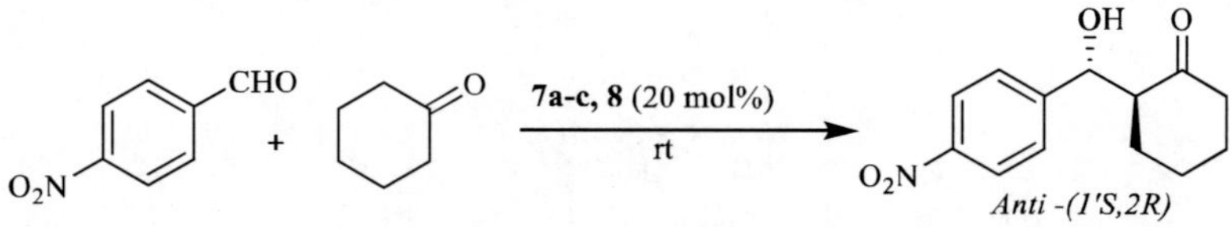
CHO
O
7a-c, 8 (20 mol%)
rt
O_2N
OH
O
Anti -(1'S,2R)

7a; Yield/time = 92/24
anti:syn = 90/10, *ee* = 54

7b; Yield/time = 91/25
anti:syn = 80/20, *ee* = 12

7c; Yield/time = 97/24
anti:syn = 90/10, *ee* = 79

8; Yield/time = 95/24
anti:syn = 98/2, *ee* = 98

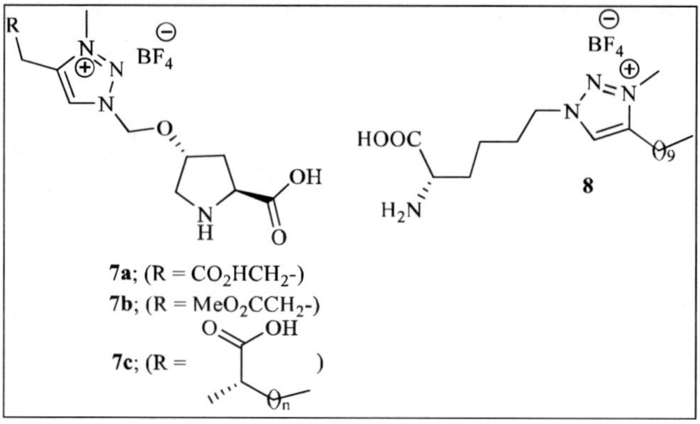
R
BF_4
OH
HOOC
H_2N
8
7a; (R = CO_2HCH_2-)
7b; (R = MeO_2CCH_2-)
7c; (R =)

Scheme 3. Direct aldol reaction catalyzed by triazolium tagged amino acid.

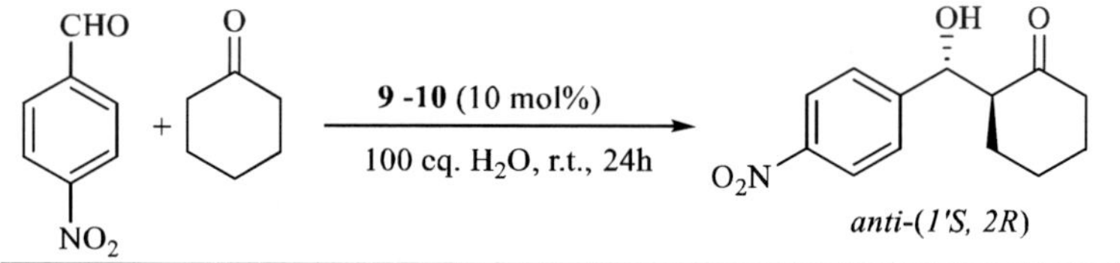
CHO
NO_2
9 -10 (10 mol%)
100 eq. H_2O, r.t., 24h
O_2N
OH
O
anti-(1'S, 2R)

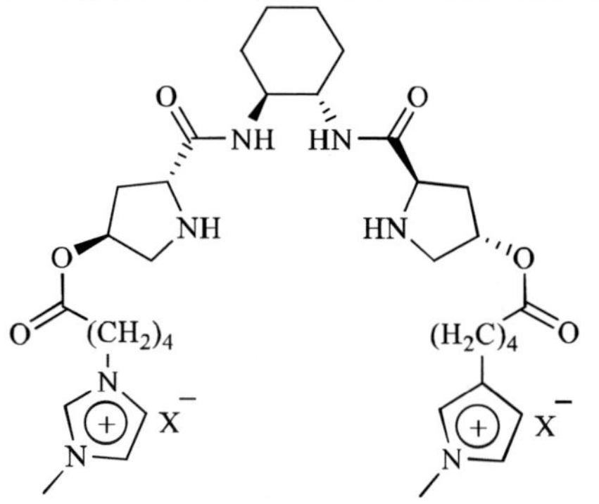
NH
HN
$(CH_2)_4$
$(H_2C)_4$
X^-

9, X = Br;
Yield = 48% *ee* = 97
anti:syn = 84:16

10, X = PF_6;
Yield = >99% *ee* = 98
anti:syn = 90:10

Scheme 4. Ionic liquid tagged prolinamide catalyzed aldol reaction.

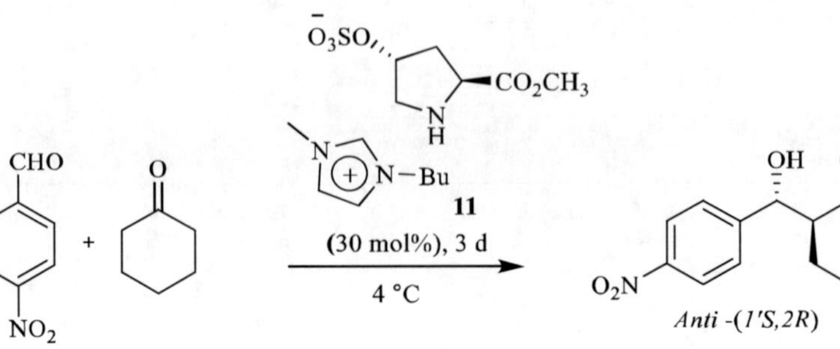

$^{-}O_3SO$ CO_2CH_3 N H N N−Bu **11**

CHO NO_2 + O (30 mol%), 3 d 4 °C OH O O_2N *Anti* -(*1'S,2R*)

[Bmim]BF_4
Anti -(*1'S,2R*), yield = 37%
anti:syn = 66:34, *ee* = 46%

[Bmim]PF_6
Anti -(*1'S,2R*), yield = 51%
anti:syn = 76:24, *ee* = 60%

[Bmim]NTf_2
Anti -(*1'S,2R*), yield = 83%
anti:syn = 96:14, *ee* = 86%

MeOH
Anti -(*1'S,2R*), yield = 53%
anti:syn = 62:38, *ee* = 22%

DCM
Anti -(*1'S,2R*), yield = 41%
anti:syn = 75:25, *ee* = 40%

Neat
Anti -(*1'S,2R*), yield = 82%
anti:syn = 59:41, *ee* = 44%

Scheme 5. Chiral imidazolium salt derived from *trans-L*-hydroxy-proline catalyzed aldol reaction.

Gauchot and Schmitzer evaluated the chiral imidazolium salt derived from *trans-L*-hydroxy-proline (11) for the direct asymmetric aldol reaction of variety of aromatic aldehydes and cyclohexanone by using different solvents like [Bmim]BF_4, [Bmim]PF_6, [Bmim]NTf_2, MeOH, CH_2Cl_2, and neat condition. ILs gave better enantioselectivities as compared to organic solvents. The ILs [Bmim]BF_4, [Bmim]PF_6, and [Bmim]NTf_2 were found to be best for obtaining good yields and stereoselectivities of the aldol product. Better results were obtained with [Bmim]NTf_2 and [Bmim]PF_6 compared to [Bmim]BF_4 but PF_6-derived ILs release the HF, [36]. The author explored the NTf_2-derived IL as a solvent for the recovery of the catalyst 11 and the catalyst was recovered and reused up to five cycles with slight loss of activity and enantioselectivity (Scheme 5) [37].

Zhang et al. designed an ionic liquid supported on polyvinylidene chloride (PVDC) (12)/*L*-proline by using simple experimental procedures. The PVDC (12)/*L*-proline is efficient for direct asymmetric aldol reaction of 4-nitrobenzaldehyde and cyclohexanone and afforded aldol adduct with excellent yield and good diastereoselectivity and enantioselectivity. The scope and limitations of PVDC ionic liquid-tagged (12)/proline organocatalytic system were also studied using a variety of benzaldehydes and cyclohexanone and the corresponding aldol products were obtained in excellent yields (up to 99%) with high diastereoselectivities of *anti* product (*dr* >6:94) with >98% *ee's*. The catalytic system efficiently recovered, recycled and retained their activity and enantioselectivity even after six cycles (Scheme 6) [38].

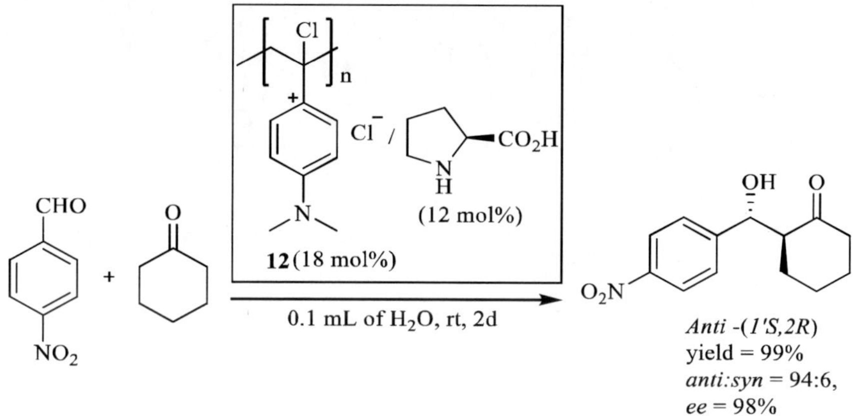

Scheme 6. Direct aldol reaction in the presence ionic liquid supported polyvinylidene chloride/*L*-proline.

Singh and co-workers reported the *trans*-4-hydroxy-(*L*)-prolinamide (13) as an organocatalyst for aldol reaction between aromatic aldehydes and cycloalkanone under solvent-free conditions [39]. Authors also designed a new hydrophobic ionic liquid 14 from bromoester of *trans*-4-hydroxy-(*L*)-prolinamide and *N*-methylimidazole, the exchange of bromide generated CILs 15 and 16 having PF_6 and BF_4 anions, respectively. The CIL 16 (2 mol%) was found to be an excellent organocatalyst for the direct asymmetric aldol reaction between 4-nitrobenzaldehyde and cyclohexanone in the presence of acetic acid (2 mol%) as cocatalyst at −15°C in solvent-free condition, and afforded corresponding aldol product in excellent yield with diastereomeric ratio (*anti*/*syn*; 97 : 3) and 94% *ee* of *anti*-aldol product (Scheme 7) [40]. (*L*)-Prolinamide imidazolium hexafluorophosphate ionic liquid 16 was reused up to 6 continuous cycles without a decrease in the conversion of the aldol product with 92% *ee* and found to be superior to its counterpart *trans*-4-hydroxy-(*L*)-prolinamide 13. Continuous cycle experiments do not require the isolation of the catalyst after each cycle. The results of reusability of the CIL 16 as catalyst were found to be better than most of the other reported reusable catalysts.

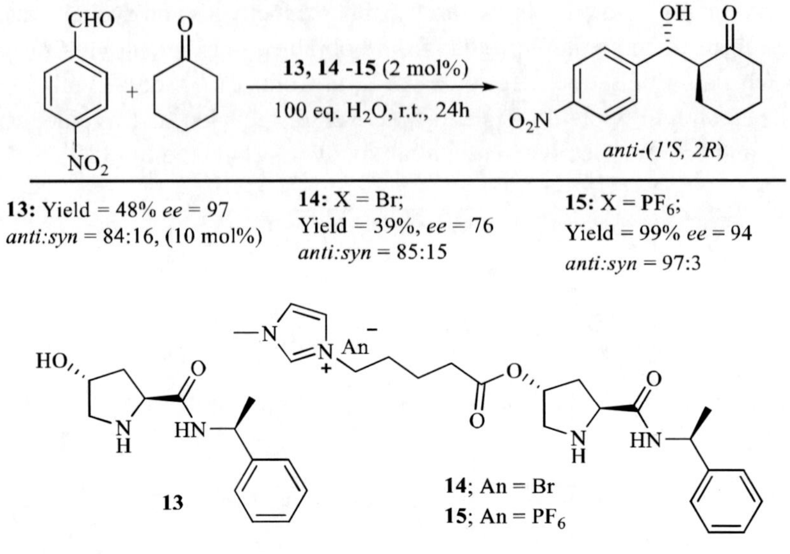

CHO O OH O

13, 14 -15 (2 mol%)

100 eq. H_2O, r.t., 24h

NO_2 O_2N

anti-(1'S, 2R)

13: Yield = 48% *ee* = 97
anti:syn = 84:16, (10 mol%)

14: X = Br;
Yield = 39%, *ee* = 76
anti:syn = 85:15

15: X = PF_6;
Yield = 99% *ee* = 94
anti:syn = 97:3

HO O N An⁻ O O O HN HN N H N H

13

14; An = Br
15; An = PF_6

Scheme 7. Prolinamide catalysts for direct asymmetric aldol reaction

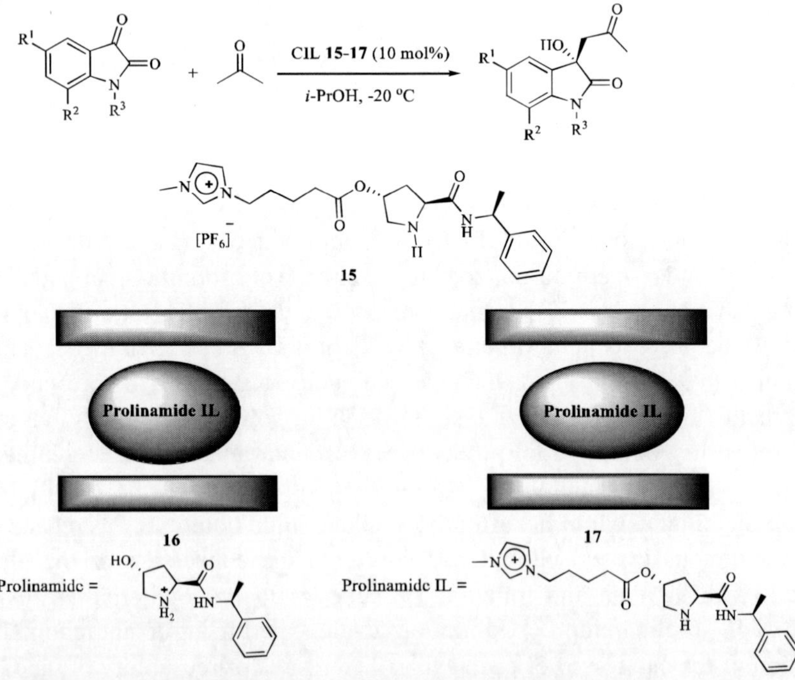

Scheme 8. Direct asymmetric aldol reaction of isatin and acetone by using prolinamide intercalated bentonite clay.

Deepa et al. developed prolinamide and imidazolium ionic liquid tagged prolinamide and were intercalated in bentonite clay and used in the asymmetric aldol reaction of isatin with acetone or cyclohexanone (Scheme 8). Chiral ionic liquid 16 (10 mol%) containing *S*-phenylethylamine gave 3-alkyl-3-hydroxy indolin-2-ones in 72-98% yields and 37-80% *ee*'s in *i*-PrOH at -20°C. The corresponding aldol products were obtained in 40-69% yields and 39-88% *ee*'s by using (10 mol%) prolinamide intercalated in bentonite clay 17. The authors found that the prolinamide intercalated bentonite clay is less reactive as compared to chiral ionic liquids containing prolinamide. The prolinamide intercalated bentonite clay 16 was recycled and reused in four cycles, with a slight loss of yield and enantioselectivity of the catalyst after the third cycle [41].

1.3. Ionic Liquid Tagged Organocatalyst Catalyzed Asymmetric Michael Addition Reaction

Over the past few years, asymmetric C-C bond-formation reactions catalyzed by organocatalysts have attracted the attention of researchers [42]. The Michael addition reaction is a conjugate addition of ketones to nitroolefins signifies an efficient method for the formation of a γ-nitrocarbonyl compound having two connecting stereocenters in a single step [43]. These nitro compounds are versatile synthetic building blocks and easily convertible into other valuable functional groups [44]. As a result, the researcher is devoted to developing efficient asymmetric catalysts, such as proline and pyrrolidine-based derivatives. These catalysts particularly environmentally benign metal-free organocatalytic systems and shown to be an efficient asymmetric catalyst for the Michael addition of aldehydes/ketones to nitroolefins [45, 46]. However, these organocatalysts still require high catalyst loading (10-30 mol%) and excess of ketone (generally 10-20 equiv.) and require low temperature which increases the limits of their application in the field of the pharmaceutical industry [47].

Therefore, researchers aim to overcome these issues by designing highly active organocatalysts have proven to be a significantly challenging task and limited success has been achieved recently [48].

Salunkhe and co-workers explored ionic liquid as a solvent and (*L*)-proline 18 for the Michael addition reaction of cyclohexanone to β-nitro styrene (Scheme 9) [49]. Authors have studied both hydrophilic and hydrophobic ILs such as 1-methoxyethyl-3-methylimidazolium methanesulphonate [MOEMIM]OMs, 1-butyl-3-methylimidazolium chloride [Bmim]Cl, 1-butyl-3-methylimidazolium tetrafluoroborate [Bmim]BF_4, 1-butyl-3-methylimidazolium hexafluorophosphate [Bmim]PF_6, for the Michael addition reaction (Scheme 9). These ionic liquids show a remarkable effect on the enantioselectivity of the reaction. The ionic liquid [MOEMIM]OMs were found to be better with respect to yield as well as stereoselectivity and (*L*)-proline was recovered and reused in [MOEMIM]OMs ionic liquid for up to three consecutive runs.

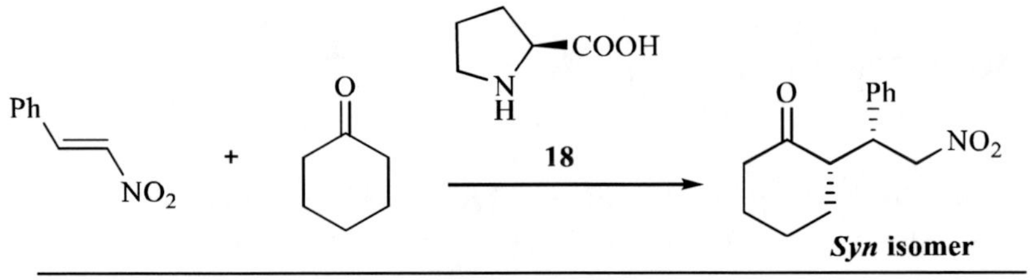

COOH
N
H
Ph
O
+
18
NO2
O Ph
NO2
***Syn* isomer**

18 (0.8 Equivalence)**; [Bmim]PF_6**; 6.5d; Yield = 68, *dr* = 90:10, *ee* = 45
18 (0.4 Equivalence)**;** [**Hmim**]**BF_4**; 6d; Yield = 53, *dr* = 60:40, *ee* = ND
18 (0.4 Equivalence)**; [MOEMIM]OMs**; 2.5d; Yield = 75, *dr* = 90:10, *ee* = 75

Scheme 9. (*L*)-Proline catalyzed Michael addition of cyclohexanone and *trans-β*-nitrostyrene in the presence of ionic liquid as the solvent.

Luo designed pyrrolidine–ionic liquids and demonstrated its activity as a highly efficient organocatalyst for the Michael addition reaction (Scheme 10) [22]. The catalytic system is efficient due to the formation of the imine bond between pyrrolidine and cyclohexanone and Michael addition to the nitroolefins and ionic liquid tagged proline not only facilitates recycling and reusability but also acts as an efficient chiral induction group to ensure high stereoselectivity. The catalytic activity and enantioselectivity varied with different ionic liquid moieties. Chiral ionic liquids with an imidazolium unit (19a-c and 21a) are superior to their 2'-methyl counterparts (20a-c and 21b) towards the activities and enantioselectivities. The protic group attached at the side chain of the cation, as in 21a and 21b, also led to a decrease in the catalytic activity and enantioselectivity.

Headley and co-workers reported the pyrrolidine sulfonamide imidazolium ILs 22-24 and IL 24 act as efficient recyclable catalyst for the Michael addition. The organocatalyst 24 showed better catalytic activity as compared to catalysts 22 and 23. These results indicate that the introduction of the sulphonyl group into the C-2 position of the electron-withdrawing imidazolium cation increases the acidity of the N-H proton of catalyst 24. The increased acidity of N-H group will help to form the stronger H-bonds during the transition states of these reactions and will avoid the use of acidic additives. The substitution of imidazolium cation increases steric hindrance closer to the catalytic site and also improves the stereoselectivity.

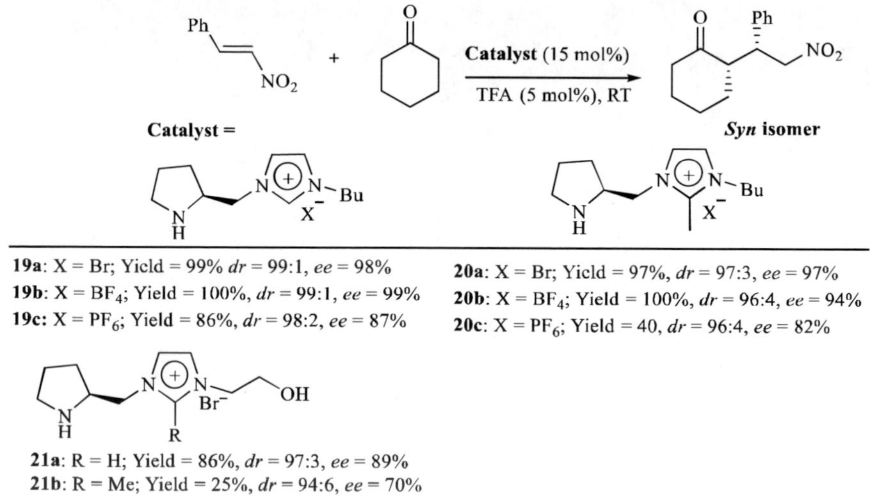

Catalyst (15 mol%)
TFA (5 mol%), RT
Syn **isomer**
Catalyst =

19a: X = Br; Yield = 99% *dr* = 99:1, *ee* = 98%
19b: X = BF_4; Yield = 100%, *dr* = 99:1, *ee* = 99%
19c: X = PF_6; Yield = 86%, *dr* = 98:2, *ee* = 87%

20a: X = Br; Yield = 97%, *dr* = 97:3, *ee* = 97%
20b: X = BF_4; Yield = 100%, *dr* = 96:4, *ee* = 94%
20c: X = PF_6; Yield = 40, *dr* = 96:4, *ee* = 82%

21a: R = H; Yield = 86%, *dr* = 97:3, *ee* = 89%
21b: R = Me; Yield = 25%, *dr* = 94:6, *ee* = 70%

Scheme 10. Michael addition reaction catalysed by ionic liquid tagged (*L*)-proline.

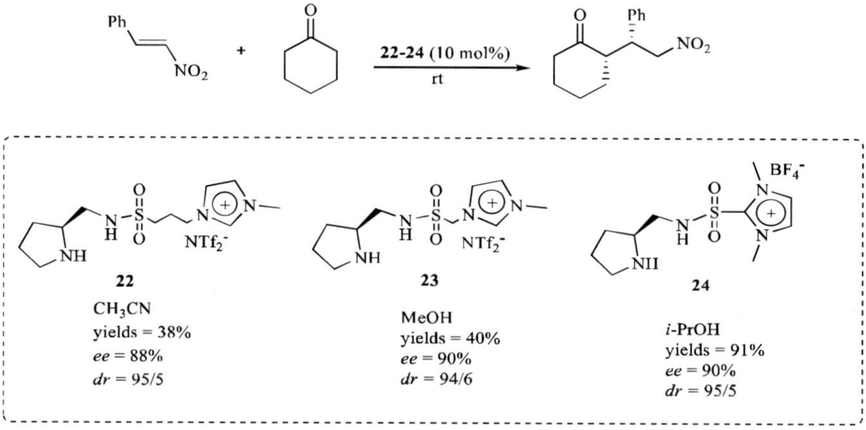

22-24 (10 mol%)
rt

22
CH_3CN
yields = 38%
ee = 88%
dr = 95/5

23
MeOH
yields = 40%
ee = 90%
dr = 94/6

24
i-PrOH
yields = 91%
ee = 90%
dr = 95/5

Scheme 11. Ionic liquid tagged sulphonamide catalyzed Michael addition reaction.

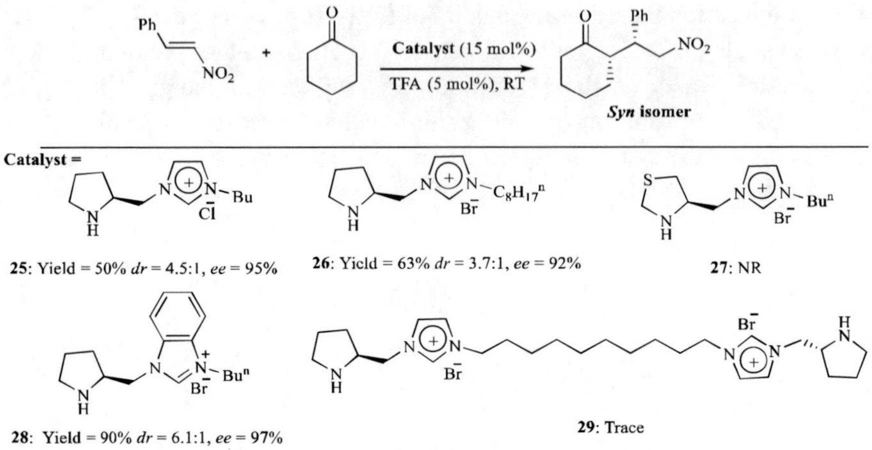

Catalyst (15 mol%)
TFA (5 mol%), RT
Syn isomer

Catalyst =

25: Yield = 50% *dr* = 4.5:1, *ee* = 95%
26: Yield = 63% *dr* = 3.7:1, *ee* = 92%
27: NR
28: Yield = 90% *dr* = 6.1:1, *ee* = 97%
29: Trace

Scheme 12. Functionalized ILs catalyzed Michael addition reaction of 4-methyl cyclohexanone and nitroolefins.

The catalyst 24 (10 mol%) acts as recyclable organocatalyst in the Michael addition reactions of *trans*-nitrostyrene and cyclohexanone as a model substrate in *i*-PrOH at room temperature and afforded corresponding product with 91% yield with 90% *ee* and 95/5 diastereoselective (Scheme 11). After completion of the reaction, the mixture was concentrated and the ethyl acetate was added for the dilution of residue and precipitation of the catalyst, which was recovered (>90%) by simply decantation of solvent. The catalyst was dried and reused for the next cycle of the reaction. Catalyst 24 was reused up to 5 times without a significant decrease in enantio and diastereoselectivity of the prodcut [50].

Cheng and co-workers synthesized functionalized chiral ILs (FCILs) and their evaluation as organocatalysts in Michael addition of 4-substituted cyclohexanone and nitrostyrene, desired Michael product bearing three stereocenters was obtained with up to 99% *ee*. Different FCILs were studied and the results are summarized in Scheme 12. The structural differences between cations show a significant role in catalytic results. The pyrrolidine-tagged FCILs showed good catalytic activity whereas the thiazolidine-containing FCIL 27 was inactive. A dication FCIL 29 was also found to be inactive for the catalysis, and the FCIL catalyst 28 containing benzimidazole moiety was found to be an efficient catalyst for the model reaction and afford the desired Michael product in 90% yield with 6.1:1 *dr* and 97% *ee* (Scheme 12) [51].

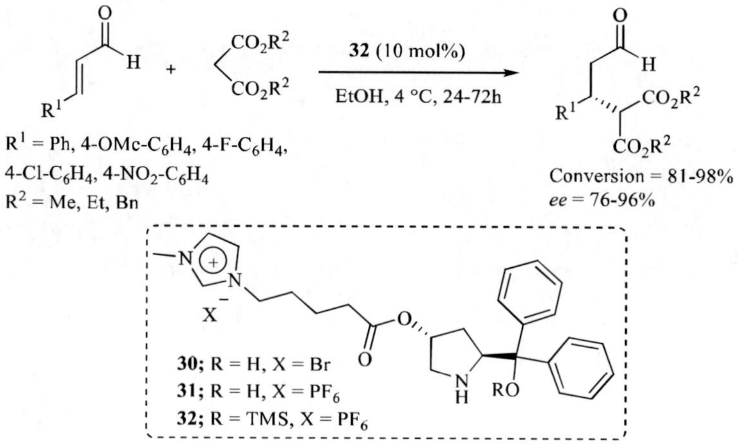

32 (10 mol%)
EtOH, 4 °C, 24-72h

R^1 = Ph, 4-OMe-C_6H_4, 4-F-C_6H_4, 4-Cl-C_6H_4, 4-NO_2-C_6H_4
R^2 = Me, Et, Bn

Conversion = 81-98%
ee = 76-96%

30; R = H, X = Br
31; R = H, X = PF_6
32; R = TMS, X = PF_6

Scheme 13. IL tagged prolinol catalyzed Michael addition reaction.

Zlotin synthesized ionic liquids tagged prolinol as recoverable organocatalysts 30-32 in 2009 and evaluated them in the Michael addition of *trans*-cinnamaldehyde and dimethyl malonate [52]. Chiral ionic liquids 30 and 31 contained free hydroxyl group, showed low catalytic activities whereas the organocatalyst 32, α -OTMS protected prolinol gave the Michael adduct in 93% yield with 96% *ee* (Scheme 13). The reaction of *trans*-cinnamaldehyde and its derivatives with dialkyl malonates afforded the corresponding products in up to 98% yield with up to 96% *ee*. The catalyst shows recyclability and reusability up to four cycles without decreases in activity.

In 2010, Zlotin also developed ionic liquid-based *cis*-prolinol organocatalysts for asymmetric Michael addition of nitroalkane and α, β-unsaturated aldehydes [53]. The silent feather of this catalyst is, that it produced (*S*)- or (*R*)-Michael products in excellent yields >99% with enantioselectivity up to 94% *ee* by changing the configuration of proline (Scheme 14). The (*R*) enantiomer of the Michael products was obtained with the *cis*-catalyst 33 and 34. Furthermore, the *trans*-isomer 32 gave almost similar conversion and enantioselectivity with opposite enantiomer (*S*) for this reaction. These ionic liquid organocatalyst was recycled up to five times without loss of yields and enantioselectivities of the product.

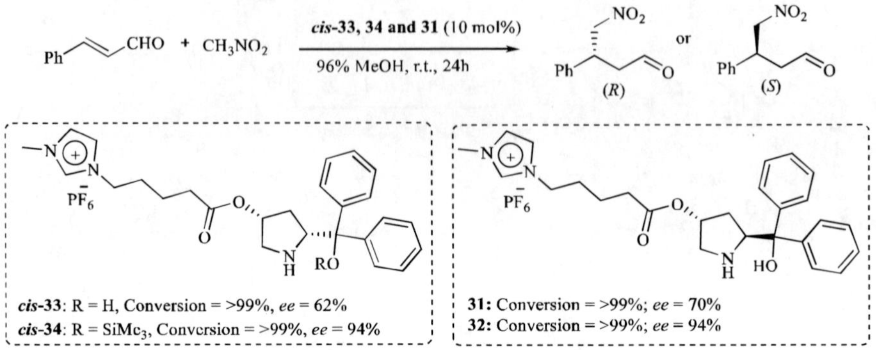

cis-33, 34 and 31 (10 mol%)

96% MeOH, r.t., 24h

cis-33: R = H, Conversion = >99%, *ee* = 62%
cis-34: R = $SiMe_3$, Conversion = >99%, *ee* = 94%

31: Conversion = >99%; *ee* = 70%
32: Conversion = >99%; *ee* = 94%

Scheme 14. IL tagged *cis*-prolinol catalyzed Michael addition reaction.

1.4. Ionic Liquids Tagged MacMillan Catalyzed Diels Alder Reaction

Otto Diels and Kurt Alder discovered the [4+2] Diels-Alder cycloaddition reaction in 1928 [54]. The [4+2] cycloaddition reaction gave monoadduct and diadduct (38a and 38b, Scheme 15) by the reaction of cyclopentadiene with quinone (37), which plays the vital role in the area of organic synthesis. The [4+2] cycloaddition reaction is one of the greatest controlling methods for the preparation of C-C bonds, which is a useful key step in the synthesis of pharmaceutical compounds and natural products [55-58].

Korolev and Mur reported the first asymmetric [4+2] cycloaddition reaction in 1948 by using enantiopure menthol [59]. Walborsky and co-workers reported [4+2] cycloaddition reaction in 1963 by the reaction of (−)-dimenthyl fumarate with butadiene in the presence of Lewis acid [60], later on, it was significantly investigated [61]. In the presence of Lewis acid, [4+2] cycloaddition reaction takes place significantly fast; Lewis's acid activates the dienophile. MacMillan described the first organocatalyzed enantioselective [4+2] cycloaddition reaction, between diene and dienophile in the presence of organocatalyst 40 to give the desire product with 82% yield and *exo:endo* (1:14) and 94% *ee* of *endo* product in MeOH/H_2O (Scheme 16) [62]. MacMillan catalyst 40 also catalyzed the [4+2] cycloaddition reaction of a variety of the substrates and gave up to 96% *ee*. The reaction mechanism was also proposed for the [4+2] cycloaddition reaction. The formation of iminium ion (III) between organocatalyst 40 and α, β unsaturated aldehyde (dienophile) and followed by *Si* attacks on cyclohexadiene (diene) to give cycloaddition product with the regeneration of organocatalyst 40 (Scheme 16).

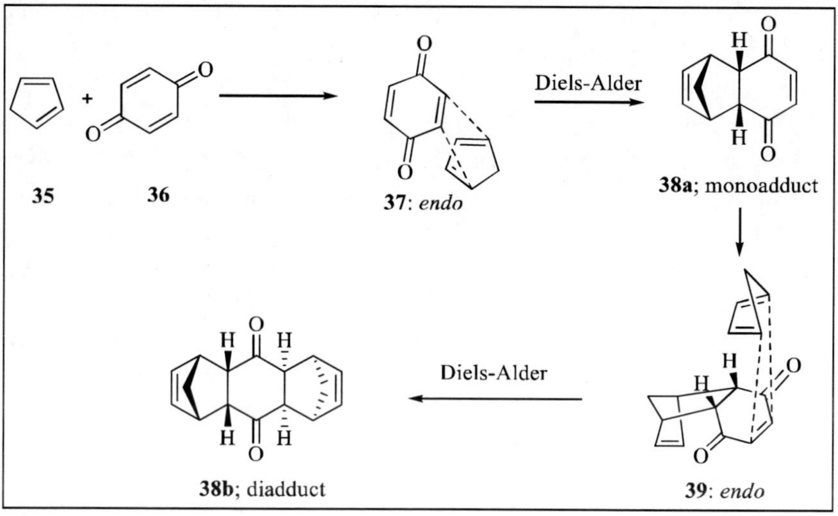

Scheme 15. [4+2] Diels-Alder reaction.

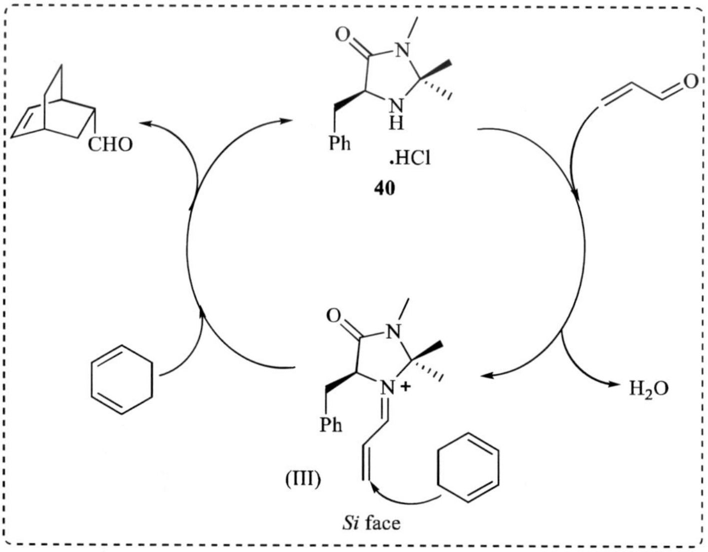

Scheme 16. The mechanism for MacMillan 40 catalyzed [4+2] cycloaddition reaction.

Park et al. used ionic liquids in asymmetric [4+2] cycloaddition reaction of cyclohexadiene and acrolein [63]. They synthesized various ionic liquids derived from the imidazole core unit by changing the anionic counterpart, like [Bmim]BF_4 41a [Bmim]PF_6 41b, [Bmim]SbF_6 41c and [Bmim]OTf 41d. The organocatalyst 40 (5 mol%) efficiently catalysed the [4+2] cycloaddition reaction between cyclohexadiene and acrolein in these ionic liquids. The reaction was monitored by TLC, after completion of the reaction diethyl ether was added to it, and the product was obtained by the decantation of the solvent layer and concentrated by evaporation while the catalyst and ionic liquid remained in the reaction vessels. The mixture of [Bmim]PF_6 IL and organocatalyst 40 was recycled and reused for the next catalytic cycle. Only 5% and 7% isolated yields were obtained with [Bmim]BF_4 and [Bmim]OTf, and *endo*-isomer was racemic (Scheme 17). It is due to the hydrophilic character of the ionic liquid, and it shows stronger H- bonding with water compared to the other two ionic liquids. The activity and enantioselectivity of the reaction were affected by water in the presence of [Bmim]OTf 41d. The transient iminium ion is hydrolysed by using at least 20% of water by volume in polar ionic liquids and enabling the catalytic cycle to turn over. The ionic liquid [Bmim]PF_6 41b and organocatalyst 40 efficiently recycled without a significant loss in activity and enantioselectivities of the product.

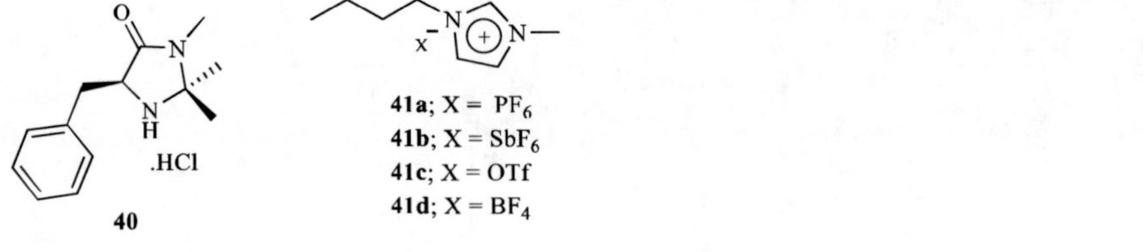

40 (5 mol%)

ionic liquid **41a-d**/H_2O = 95/5 v/v

10 °C, 18 h

CHO

41a; Yield = 76%, *endo:exo* = 17:1, *ee* of *endo* = 93%

41b; Yield = 74%, *endo:exo* = 17:1, *ee* of *endo* = 92%

41c; Yield = 7%, *endo:exo* = 17:1, *ee* of *endo* = 0%

41d; Yield = 5%, *endo:exo* = 17:1, *ee* of *endo* = 0%

.HCl

40

41a; X = PF_6

41b; X = SbF_6

41c; X = OTf

41d; X = BF_4

Scheme 17. Diels-Alder reaction between cyclohexadiene and acrolein in the presence of MacMillan catalyst and ionic liquids.

Diels–Alder cycloaddition reaction cataylezd by MacMillan catalyst 40 in an ionic liquid/H_2O homogeneous phase was investigated by Nino and co-workers [64]. The [Mpy(OTf)] or [Bmim(OTf) as ionic liquids with

MacMillan catalyst 40. HCl have been investigated in this reaction (Scheme 18). This catalytic system found to be superior in terms of activity, selectivity, recyclability, and required less reaction time as compare to traditional several organic solvents. [Mpy(OTf)]- IL 42 gave excellent activity with catalyst 40 than the [Bmim(OTf)] IL in the [4+2] cycloaddition reaction. The [4+2] cycloaddition reaction of cyclohexadiene and acrolein by using catalyst 40·HCl/[Mpy(OTf)] gave desired product with 98% yield and 93% *ee* of the *endo* product and organocatalyst 39·HCl/[Bmim(OTf)] 41c gave 97% yield and 88% *ee* of *endo* product (Scheme 18). They have concluded that the [4+2] cycloaddition has also catalyzed with less catalyst loading (1 mol%) without decreasing the *endo*:*exo* ratio with a slight reduction of enantiomeric excess. The water is the effective co-solvent in this reaction and allowing it to a homogeneous stage when the reaction is carried out in the presence of [Mpy(OTf)] 42, and helping the hydrolysis of iminium ion. When increasing the ring size from 5 to 7 in the substrate also increased the diastereomeric ratio of *endo*-product.

The ionic liquid-supported imidazolidinone catalysts 43a-c were synthesized by Shen and evaluated in an enantioselective Diels-Alder reaction [65]. The organocatalysts 43a-c (10 mol%) and trifluoroacetic acid (10 mol%) as co-catalyst in CH_3CN/H_2O at room temperature were found to be the best catalytic system for the [4+2] cycloaddition reaction and provide cycloaddition product in 97% yield with 88% *ee* for *exo* product and 93% *ee* for *endo* product. The imidazolium-based ionic liquid-tagged MacMillan catalyst 43a was efficiently recycled and reused up to 5 catalytic cycles without loss of activity and selectivity (Scheme 19).

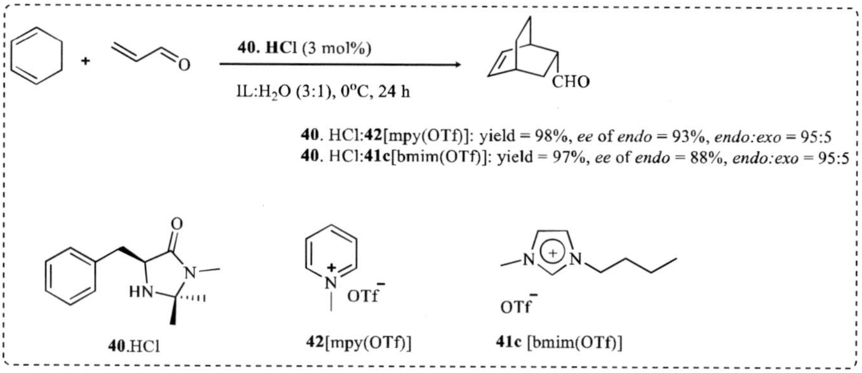

40. HCl (3 mol%)
IL:H_2O (3:1), 0°C, 24 h
CHO

40. HCl:**42**[mpy(OTf)]: yield = 98%, *ee* of *endo* = 93%, *endo:exo* = 95:5
40. HCl:**41c**[bmim(OTf)]: yield = 97%, *ee* of *endo* = 88%, *endo:exo* = 95:5

40.HCl **42**[mpy(OTf)] **41c** [bmim(OTf)]

Scheme 18. Ionic liquids and MacMillan catalyzed Diels-Alder reaction between cyclohexadiene and acrolein.

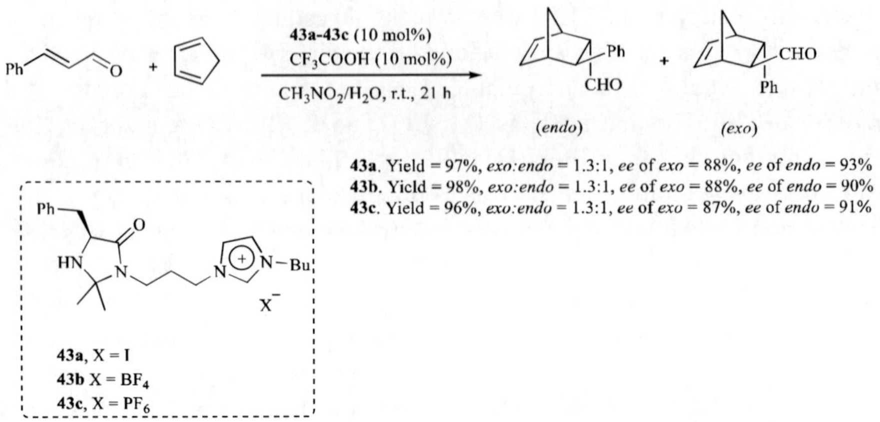

43a-43c (10 mol%)
CF_3COOH (10 mol%)
CH_3NO_2/H_2O, r.t., 21 h

(*endo*) (*exo*)

43a. Yield = 97%, *exo:endo* = 1.3:1, *ee* of *exo* = 88%, *ee* of *endo* = 93%
43b. Yield = 98%, *exo:endo* = 1.3:1, *ee* of *exo* = 88%, *ee* of *endo* = 90%
43c. Yield = 96%, *exo:endo* = 1.3:1, *ee* of *exo* = 87%, *ee* of *endo* = 91%

43a, X = I
43b X = BF_4
43c, X = PF_6

Scheme 19. Ionic liquid tagged imidazolidinones catalyzed Diels-Alder reaction between cyclopentadiene and *α, β*-unsaturated aldehydes.

44 (6 mol%)
[mpy(OTf)]:HCl (3:1), 0°C, 16 h

endo

yield = 98%; *ee* of *endo* = 90%
endo:exo = 98:2

44

42 Ionic liquid [mpy(OTf)]

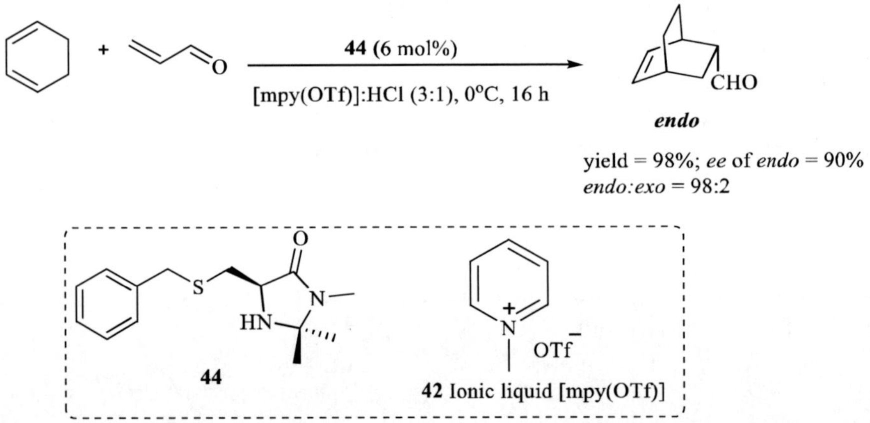

Scheme 20. [4+2] Cycloaddition reaction between cyclohexadiene and *α, β*-unsaturated aldehydes in the presence of ionic liquid.

Author also studies the [4+2] cycloaddition reaction between dienes and *α, β*-unsaturated aldehydes by using sulfur atom containing imidazolidinone 44 and [Mpy(OTf)] ionic liquid. The phenyl group of imidazolidinone bearing sulfur atom 44 allow to attacks the less hindered *Si*-face of dienophiles which is shown by the conformational study. The 6 mol% of organocatalyst 44 in Mpy[OTf]/HCl (3/1) best catalytic system for the [4+2] cycloaddition reaction of diene and acrolein and afforded the cycloaddition product in 98% yield and diastereomeric ratio (*endo:exo*; 98:2) and 90% *ee* of *endo* isomer at 0°C

(Scheme 20). The organocatalyst 44 in [Mpy(OTf)] ionic liquid 42 was efficiently recycled and reused for 6 catalytic cycles without loss of yield and enantioselectivity [66].

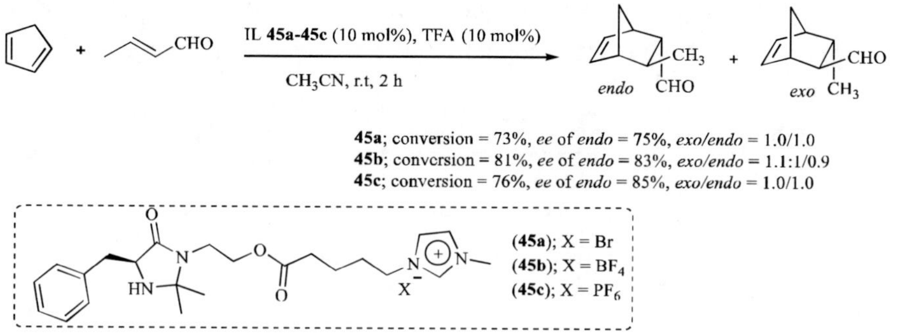

IL **45a-45c** (10 mol%), TFA (10 mol%)

CH_3CN, r.t, 2 h

endo *exo*

45a; conversion = 73%, *ee* of *endo* = 75%, *exo*/*endo* = 1.0/1.0
45b; conversion = 81%, *ee* of *endo* = 83%, *exo*/*endo* = 1.1:1/0.9
45c; conversion = 76%, *ee* of *endo* = 85%, *exo*/*endo* = 1.0/1.0

(**45a**); X = Br
(**45b**); X = BF_4
(**45c**); X = PF_6

Scheme 21. Modified MacMillan ionic liquids catalyzed [4+2] cycloaddition reaction.

The ionic liquids tagged modified MacMillan catalysts through an ester linkage were synthesized by Singh and co-workers in 2015 and these recyclable and reusable organocatalysts 45a-c were evaluated in asymmetric [4+2] cycloaddition reaction [67]. The [4+2] cycloaddition reaction between diene and crotonaldehyde in the presence of organocatalyst 45 (10 mol%) and TFA (10 mol%) to obtain corresponding product in 76% conversion and 85% *ee* in acetonitrile. After optimization of reaction conditions, 5 mol% of organocatalyst 45c and 5 mol% of trifluoroacetic acid (TFA) as a co-catalyst in CH_3CN/H_2O (95:5) were found to be suitable conditions in substrate scope and afforded corresponding product in up to 84% yield with diastereomeric ratio (1/1.1, *exo/endo*) and up to 90% enantiomeric excess of major product (*endo* isomer) at 25°C temperature. The ionic liquid tagged modified MacMillan organocatalyst 45c was recycled and reused for 5 catalytic cycles with a minor decrease in conversion and enantioselectivity of the [4+2] cycloaddition product (Scheme 21).

Deepa et al. reported modified MacMillan organocatalyst 46 and ionic liquids of dicationic imidazolium (47a-47c) used as catalysts for asymmetric Diels-Alder reaction [68]. The [4+2] cycloaddition reaction of cyclopentadiene and crotonaldehyde by using modified MacMillan organocatalyst 46 (5 mol%), IL 47a-c (10 mol%) and trifluoroacetic acid (TFA) (5 mol%) as a co-catalyst in CH_3CN/H_2O (95/5), the Diels-Alder adduct was obtained in 95% conversion with 87% *ee* of *endo* isomer (Scheme 22). The modified MacMillan organocatalyst was shown better recovery when 20 mol% dicationic imidazolium IL 47c was used. The catalyst was recycled up to 5 cycles without loss in conversion of the product and slightly dropped in *ee* of the *endo* product. The dicationic ionic liquid was found to be efficient in recovery of the catalyst 45 compared it monocationic ionic liquid.

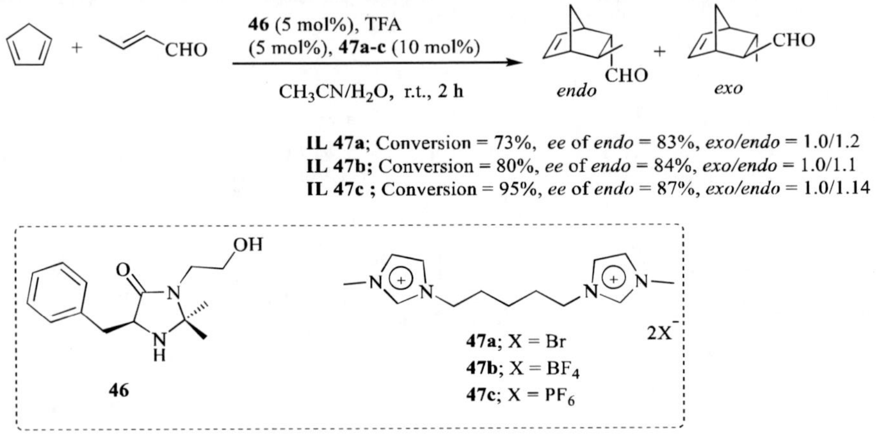

46 (5 mol%), TFA (5 mol%), **47a-c** (10 mol%)

CH_3CN/H_2O, r.t., 2 h

CHO; *endo*; *exo*

IL 47a; Conversion = 73%, *ee* of *endo* = 83%, *exo/endo* = 1.0/1.2
IL 47b; Conversion = 80%, *ee* of *endo* = 84%, *exo/endo* = 1.0/1.1
IL 47c ; Conversion = 95%, *ee* of *endo* = 87%, *exo/endo* = 1.0/1.14

46

47a; X = Br
47b; X = BF_4
47c; X = PF_6

Scheme 22. Modified MacMillan catalyst and dicationic ionic liquids catalyzed asymmetric Diels-Alder reaction.

1.5. Ionic Liquid Tagged Organocatalyst Catalyzed Asymmetric Reduction of Carbonyls

The chiral secondary alcohols are important class of organic compounds which is formed by enantioselective reduction of prochiral ketones [69, 70]. These alcohols are chiral building blocks and used for the synthesis of pharmaceuticals, natural products, and agrochemicals. Hirao developed the most popular homogeneous chiral 1,3,2- oxazaborolidine catalyst and BH_3. THF catalyzed asymmetric reduction of prochiral ketones at room temperature and higher temperature [71-79]. The homogeneous chiral 1,3,2-oxazaborolidines were difficult to separate from the reaction mixture because of their low molecular weight. To overcome these issues researchers modified the catalyst with fluorous tags [80], imidazolium-tagged sulfonamide [81], triazole linked dendrimers [82], use of C_3-symmetric tris(β-hydroxyphosphoramide) [52], and immobilization on polymer beads by the covalent bond [83–86].

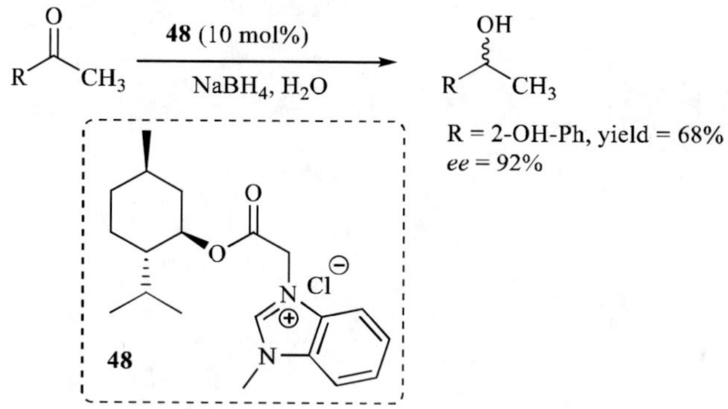

Scheme 23. Asymmetric reduction of prochiral ketones using ionic liquids tagged modified *α, α*-diphenyl-4-hydroxy-(*L*)-prolinol.

Chopra and co-workers reported enantioselective reduction of prochiral ketones by using benzimidazolium salt-based ionic liquid and sodium borohydride as a hydride source (Scheme 23) with enantioselectivities up to 90% and moderate yields. The thermal stability and fluorescence activity are also shown by chiral IL 48. The catalyst 48 was found to be efficient with respect to activity (68% yield) and enantioselectivity (92% *ee*). Catalyst 48 generates a 1,2,3-oxazaborolidine complex with BH_3. SMe_2 which is the active catalyst for the reduction of ketones [87].

Singh and co-workers synthesized ionic liquids tagged modified catalyst 30-31 and 49-50 from the reaction of *trans-α, α*-diphenyl-4-hydroxy-(*L*)-prolinol, 5-bromovaleric acid and imidazole. The *in-situ* generated 1,3,2-oxazaborolidine species from the reaction of $BH_3 \cdot SMe_2$ with ionic liquid tagged *α, α*-diphenyl-4-hydroxy-(*L*)-prolinol, which acts as an active catalyst for asymmetric reduction of prochiral ketones. The 10 mol% of ionic liquid tagged *α, α*-diphenyl-4-hydroxy-(*L*)-prolinol hexafluorophosphate (PF_6) anions 31 and $BH_3 \cdot SMe_2$ was used in the reduction of acetophenone, afforded 1-phenylethanol in 99% yield with 84% *ee* (Scheme 24) [88]. This catalytic system is also efficient for a variety of substrates. Based on further studies during the recyclability of the reaction, the IL 50 with ether linkage was found to be stable as compared to the IL 49 with ester linkage against the BH_3. SMe_2. The *in-situ* generated 1,3,2-oxazaborolidine species of IL 50 was reused and recycled up to 4 cycles with minor loss of catalytic activity and enantioselectivity.

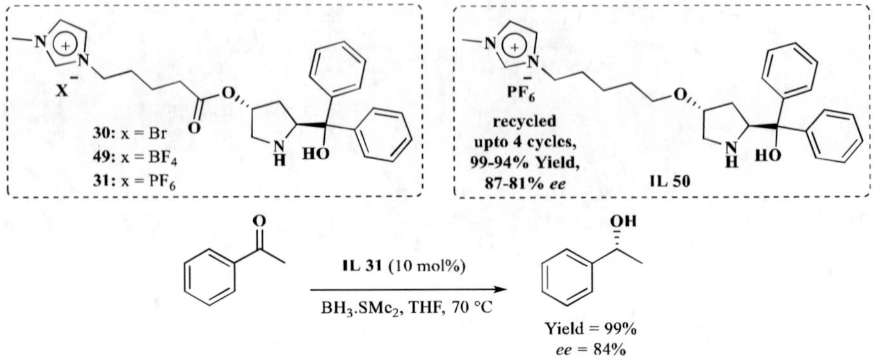

Scheme 24. Asymmetric reduction of prochiral ketones using ionic liquids tagged modified *α, α*-diphenyl-4-hydroxy-(*L*)-prolinol.

1.6. Ionic Liquid Tagged Organocatalyst Catalyzed Asymmetric Friedel-Crafts Alkylation

The Friedel-Crafts alkylation reaction is one of the most powerful methods for the preparation of C-C bond in the organic synthesis [89]. The organocatalyzed asymmetric F-C reaction received more attention of researchers after discovery of aldol and Diels-Alder reactions[90]. The first-generation MacMillan catalyst has been reported by MacMillan in asymmetric F-C alkylation reaction of pyrrole and *α, β*-unsaturated aldehyde [85]. The 4-imidazolidin-4-one containing second-generation organocatalyst with additional stereocenter also developed by MacMillan and used in asymmetric enantioselective F-C alkylation reaction of indoles and *α, β*-unsaturated aldehyde [91]. The indole structural motif is found in a wide range of natural products and shows various pharmaceutical properties [92]. The asymmetric F-C alkylation reaction has been reported by researchers using various organocatalysts *viz*: chiral imidazolium-4-one [93], imidazolethiones [94], chiral cinchona alkaloids [95], chiral phosphorylated imidazolidiones [96], aziridine-2-yl methanols [97], bifunctional squaramides [98], chiral diarylprolinol trimethylsilyl ethers [99], and cited in review article [100].

The second-generation MacMillan catalyst modified with imidazolium IL has been synthesized by Jubair et al. and used as an organocatalyst (10 mol%) and trifluoroacitic acid (10 mol%) as an additive at −60°C in asymmetric F-C alkylated reaction of *N*-benzylindole and *α, β*-unsaturated aldehydes and the desire product was obtained in 87% yield with 89% *ee*. Second-generation

MacMillan catalyst tagged with IL was also an efficient organocatalyst for substituted indoles and a variety of *α, β*-unsaturated aldehydes and gave 43-95% yields and 58-90% *ee*'s of alkylated indoles (Scheme 25). The second-generation organocatalyst was recycled and reused up to 4 cycles and substantial drop in yields and *ee* was observed after third cycle [101].

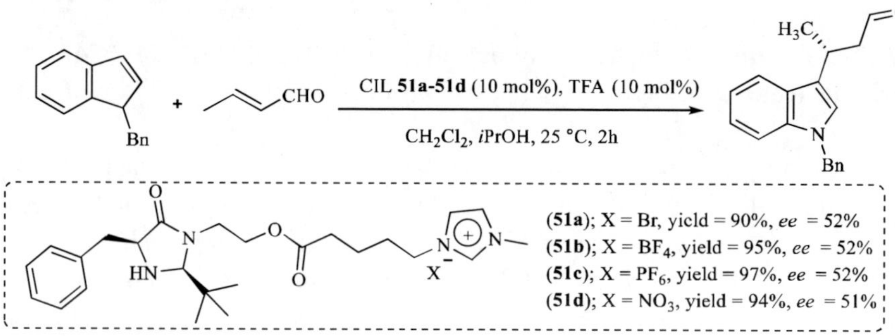

CIL **51a-51d** (10 mol%), TFA (10 mol%)

CH_2Cl_2, *i*PrOH, 25 °C, 2h

(**51a**); X = Br, yield = 90%, *ee* = 52%
(**51b**); X = BF_4, yield = 95%, *ee* = 52%
(**51c**); X = PF_6, yield = 97%, *ee* = 52%
(**51d**); X = NO_3, yield = 94%, *ee* = 51%

Scheme 25. The second-generation MacMillan modified with imidazolium IL catalyzed asymmetric F-C reaction.

CIL **31, 32, 52a-b, 53a-b** (20 mol%)
Et_3N (20 mol%), 1,4-dioxane, 25 °C, 24h
ii) $NaBH_4$, MeOH, 0 °C, 20min

(**32**); Ar = Ph, n = 4, yield =47%, *ee* = 72%
(**52a**); Ar = 3,5-$(CF_3)_2C_6H_3$, n = 4, yield = 52%, *ee* = 56%
(**52b**); Ar = Ph, n = 2, yield = 45%, *ee* = 70%

(**31**); Ar = Ph, n = 4, yield = 19%, *ee* = 26%
(**53a**); Ar = 3,5-$(CF_3)_2C_6H_3$, n = 4, yield = 18%, *ee* = 20%
(**53b**); Ar = Ph, n = 2, yield = 22%, *ee* = 24%

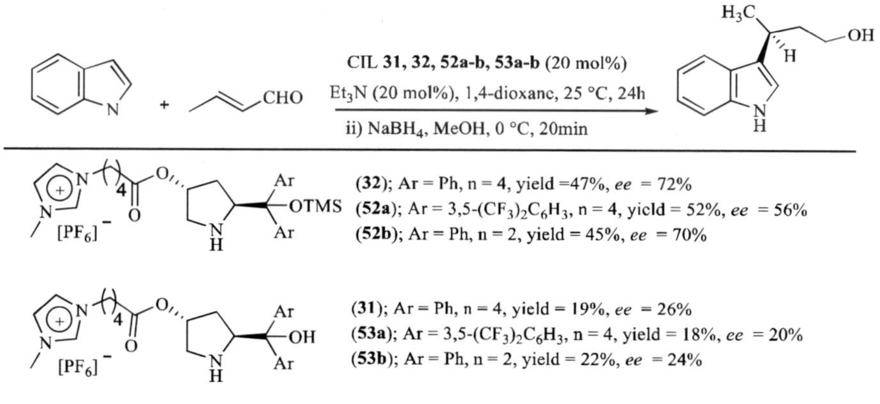

Scheme 26. *α,α*-Diaryl-(*S*)-prolinol trimethylsilyl ether tagged with imidazolium-based chiral IL catalyzed asymmetric F-C reaction.

Singh and co-workers developed *α, α*-diaryl-(*S*)-prolinol trimethylsilyl ether tagged with imidazolium-based chiral IL and used in asymmetric F-C alkylation reaction of indoles and *α, β*-unsaturated aldehydes (Scheme 26). The *α, α*-diphenyl-(*S*)-prolinol trimethyl ether tagged imidazolium hexafluorophosphate anion $[PF_6]$ (20 mol%) catalyzed F-C alkylation reaction in the presence of triethylamine (40 mol%) as an additive in 1,4-dioxane at room temperature afforded 3-alkylated indoles in 62-89% yields with 47-88% *ee's*. The organocatalyst 32 was also recycled and reused for asymmetric F-C reaction up to seven cycles without loss of enantioselectivity and considerable loss in yields of the 3-alkylated indoles after the third cycle [102].

1.7. Ionic Liquid Tagged Organocatalyst Catalyzed Asymmetric Biginelli Reaction

The Biginelli reaction is the most important multi component reaction. The Pietro Biginelli reported the three-component condensation reaction between aromatic aldehyde, urea and *β*-keto ester in 1891 [103]. 3,4-Dihydropyrimidin-2-(1*H*)-ones (DHPMs) structure motif consists of a heterocyclic system which has important pharmacological properties such as antiviral [104], antitumor [105], antibacterial [106] and anti-inflammatory [108]. Hence, researchers have great interest to accessing enantiomerically pure DHPMs by asymmetric catalysis.

Deepa et al. synthesized ionic liquid tagged (*L*)-prolinamide and used as an organocatalyst in enantioselective Biginelli reaction of benzaldehyde, urea and *β*-keto esters. The synthesized organocatalyst were characterized by FT-NMR, FT-IR, HRMS, and TGA. The (*L*)-prolinamide modified with various counter ion containing ionic liquids 14, 15, 54 and 55a-c (10 mol%) and trifluoro acetic acid (TFA) (10 mol%) as an additive efficiently catalyzed the Biginelli reaction and chiral 3,4-dihydropyrimidin-2-1*H*)-ones (DHPMs) were obtained in 19-39% yields and 34-65% *ee*'s at room temperature in 24h (Scheme 27). The adamantly (*L*)-prolinamide containing $[BF_4]^-$ as counter anion (5 mol%) and *p*-toluene sulphonic acid (*p*-TSA) (5 mol%) was found to be the best catalytic system for the enantioselective Biginelli reaction. The plausible reaction mechanism and intermediates or transition states are also proposed by the author and confirmed by FT-NMR and DFT calculations. The asymmetric induction was introduced by attacks of ethylacetoacete to *Re*-face on the iminium bond and (*R*)-absolute configuration of Biginelli products was obtained [108].

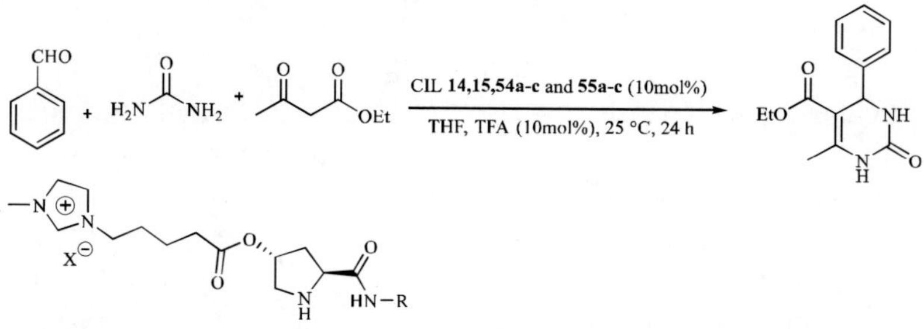

CIL **14,15,54a-c** and **55a-c** (10mol%)
THF, TFA (10mol%), 25 °C, 24 h

14: R = (*S*)-1-phenylethyl, [X] = [Br], yield = 21%, *ee* = 42%
15: R = (*S*)-1-phenylethyl, [X] = [PF_6], yield = 28%, *ee* = 34%
54: R = (*S*)-1-phenylethyl, [X] = [BF_4], yield = 39%, *ee* = 47%

55a: R = 1-Adamantyl, [X] = [Br], yield = 19%, *ee* = 52%
55b: R = 1-Adamantyl, [X] = [PF_6], yield = 26%, *ee* = 39%
55c: R = 1-Adamantyl, [X] = [BF_4], yield = 39%, *ee* = 65%

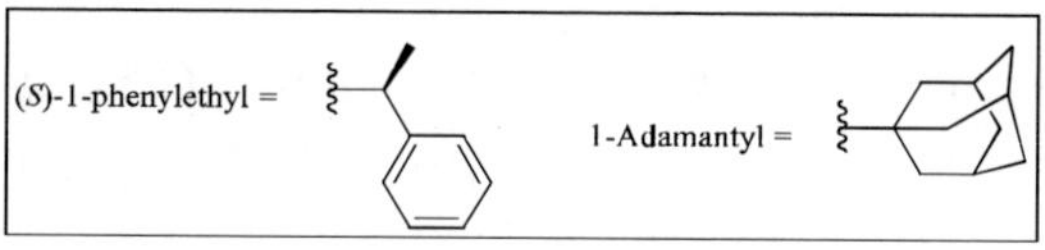

(*S*)-1-phenylethyl =
1-Adamantyl =

Scheme 27. (*L*)-Prolinamide modified with various counter ions containing ionic liquids catalyzed enantioselective Biginelli reaction.

2. Future Trends

Over the recent years, a numerous progress has been made towards the synthesis of ionic liquid and their application as an organocatalyst for the asymmetric organic transformations. The IL tagged organocatalysts are becoming practicable selections of researchers for widely interesting catalytic asymmetric organic transformations. In the instance of proline as an organocatalysts, the organocatalyst is usually used in high catalyst loading in a nonvolatile solvent like DMSO. To overcome these issues, several research groups are attentive on developing recoverable chiral ionic liquid as an organocatalyst for asymmetric organic reactions. The recovery and reusability of organocatalyst became dynamic for reducing the cost, making the process greener as the viewpoint of the environment aspect, and facilitating the separation of the product from the catalyst. Moreover, the improvement of productivity and catalyst activity is challenging from commercial applications. The investigation towards the effective cost and greener strategies will be highly advantageous for the development of asymmetric catalytic methodologies. The organocatalyst tagged with ionic liquid has much more scope for different asymmetric organic transformations.

Conclusion

In this chapter, we have discussed several libraries of ionic liquid tagged organocatalyst for the asymmetric organic transformations (including Aldol reaction, Michael addition reaction, Diels Alder reaction, asymmetric reductions of carbonyls, Friedel-Crafts alkylation reaction, asymmetric Biginelli reaction). Ionic liquids have outstanding potential as organocatalysts and as a solvent in recent years. Various organocatalysts such as ionic liquids tagged with proline, prolinamide, MacMillan catalyst, prolinol have been used in the asymmetric organic transformations. The ionic liquids are composed of ions and polar in nature therefore, insoluble in a non-polar solvent and this property makes them recoverable and reusable organocatalysts. The recovery and reuse of the organocatalyst is greener for environment aspect and dynamic for reducing the cost and facilitating the separation of the product from the catalyst. The immobilized organocatalysts provide some advantages, such as easy experimental procedures, simple reaction set-up, nontoxicity, recyclability, and reusability. Immobilization of modified organocatalysts is desirable due to their commercial feasibility and their high reactivity.

References

[1] Wasserscheid P, Keim W. Ionic Liquids New "Solutions" for Transition Metal Catalysis. *Angew. Chem. Int. Ed.*, (2000) 39: 3772-3789.

[2] Eds. Wasserscheid P, Welton T. Outlook: Ionic liquid in synthesis. *Wiley-VCH Verlag GmbH & Co. KGaA.* (2002) 348-355.

[3] Hallett JP, Welton T. Room-Temperature Ionic Liquids: Solvents for Synthesis and Catalysis. *Chem. Rev.*, (2011) 111: 3508– 3576.

[4] Bara JE, Carlisle TK, Gabriel CJ, Camper D, Finotello A, Gin DL, Nobel RD. Guide to CO_2 Separations in Imidazolium-Based Room-Temperature Ionic Liquids. *Ind. Eng. Chem. Res.* (2009) 48: 2739–2751.

[5] Lei Z, Dai C, Chen B. Gas Solubility in Ionic Liquids. *Chem. Rev.* (2014) 114: 1289–1326.

[6] Lei Z, Dai C, Zhu J, Chen B. Extractive distillation with ionic liquids: A review. *AIChE J.* (2014)60: 3312–3329.

[7] Chatel G, MacFarlane DR. Ionic liquids and ultrasound in combination: synergies and challenges. *Chem. Soc. Rev.* (2014) 43: 8132–8149.

[8] Mai NL, Koo YM, Computer-Aided Design of Ionic Liquids for High Cellulose Dissolution. *ACS Sustainable Chem. Eng.* (2016) 4: 541–547.

[9] Gurkan BE, de la Fuente J, Mindrup EM, Ficke LE, Goodrich BF, Price EA, Schneider WF, Brennecke JF. Equimolar CO_2 Absorption by Anion-Functionalized Ionic Liquids. *J. Am. Chem. Soc.* (2010) 132: 2116−2117.

[10] Ruckart KN, O'Brien RA, Woodard SM, West KN, Grant T. Porous Solids Impregnated with Task-Specific Ionic Liquids as Composite Sorbents. (2015) *J. Phys. Chem. C* 119: 20681− 20697.

[11] Qian W, Texter J, Yan F, Frontiers in poly(ionic liquid)s: syntheses and applications. *Chem. Soc. Rev.* (2017) 46: 1124−1159.

[12] Rojas MF, Bernard FL, Aquino A, Borges J, Dalla Vecchia F, Menezes S. Ligabue R. Einloft S. Poly(ionic liquid)s as efficient catalyst in transformation of CO_2 to cyclic carbonate. *J. Mol. Catal. A: Chem.* (2014) 392: 83−88.

[13] Wickramanayake S, Hopkinson D, Myers C, Hong L, Feng J, Seol Y, Plasynski D, Zeh M, Luebke D. Mechanically robust hollow fiber-supported ionic liquid membranes for CO_2 separation applications. *J. Membr. Sci.* (2014) 470: 52−59.

[14] Scovazzo P, Havard D, McShea M, Mixon S, Morgan D. Long-term, continuous mixed-gas dry fed CO_2/CH_4 and CO_2/N_2 separation performance and selectivities for room temperature ionic liquid membranes. *J. Membr. Sci.* (2009) 327: 41−48.

[15] Khan NA, Hasan Z, Jhung SH. Ionic Liquids Supported on Metal-Organic Frameworks: Remarkable Adsorbents for Adsorptive Desulfurization. *Chem. - Eur. J.* (2014) 20: 376−380.

[16] Vicent-Luna, JM, Gutierrez-Sevillano JJ, Anta JJ, Calero´ S. Effect of Room-Temperature Ionic Liquids on CO_2 Separation by a Cu-BTC Metal-Organic Framework. *J. Phys. Chem. C.* (2013) 117: 20762−20768.

[17] Welton T. Room-Temperature Ionic Liquids. Solvents for Synthesis and Catalysis *Chem. Rev.* (1999) *99*: 2071-2084.

[18] Pârvulescu VI, Hardacre C. Catalysis in Ionic Liquids. *Chem. Rev.* (2007) 107: 2615-2665.

[19] Rantwijk FV, Sheldon RA. Biocatalysis in Ionic Liquids. *Chem. Rev.* (2007) 107: 2757-2785.

[20] D. Wei, A. Ivaska. Applications of ionic liquids in electrochemical sensors *Analytica Chimica Acta*, (2008) 607: 126-135.

[21] Herrmann WA, Goossen LJ, Kocher C, Artus GRJ. Chiral heterocylic carbenes in asymmetric homogeneous catalysis. *Angew. Chem. Int. Ed. Engl.* (1996) 35: 2805–2807.

[22] Luo S, Mi X, Zhang L, Liu S, Xu H, Cheng J-P. Functionalized Chiral Ionic Liquids as Highly Efficient Asymmetric Organocatalysts for Michael Addition to Nitroolefins. *Angew. Chem. Int. Ed.* (2006) 45: 3093-3097.

[23] Earle M, McCormac P, Seddon K. Diels–Alder reactions in ionic liquids. A safe recyclable alternative to lithium perchlorate–diethyl ether mixtures. *Green Chem.* (1999) 1: 23–25.

[24] Bao W, Wang Z, Li Y. Synthesis of chiral ionic liquids from natural amino acids. *J. Org. Chem.* (2003) 68: 591–593.

[25] Fukumoto K, Yoshizawa M, Ohno H. Room temperature ionic liquids from 20 natural amino AIDS. *J. Am. Chem. Soc.* (2005) 127: 2398–2399.

[26] Gausepohl R, Buskens P, Kleinen J, Bruckmann A, Lehmann CW, Klankermayer J, Leitner W. Highly enantioselective aza-Baylis-Hillman reaction in a chiral reaction medium. *Angew. Chem. Int. Ed.* (2006) 45: 3689–3692.

[27] Branco LC, Gois PMP, Lourenco NMT, Kurteva VB, Alfonso CAM. Simple transformation of crystalline chiral natural anions to liquid medium and their use to induce chirality. *Chem. Commun.* (2006) 22: 2371–2372.

[28] Yu L, Jin X, Zeng X. Methane interactions with polyaniline /butylmethylimidazolium camphorsulfonate ionic liquid composite. *Langmuir* (2008) 24: 11631–11636.

[29] Payagala T, Armstrong DW. Chiral ionic liquids: A compendium of syntheses and applications (2005–2012). *Chirality* (2012) 24: 17–53.

[30] Karimi BK, Tavakolian M, Akbari M, Mansouri F. Ionic liquids in asymmetric synthesis: An overall view from reaction media to supported ionic liquid catalysis. *ChemCatChem* (2018) 10: 3173–3205.

[31] Chiappe C, Marra A, Mele A. Synthesis and applications of ionic liquids derived from natural sugars. *Top. Curr. Chem.* (2010) 295: 177–195.

[32] Luo S, Mi X, Zhang L, Liu S, Xu H, Cheng JP. Functionalized ionic liquids catalyzed direct aldol reactions. *Tetrahedron,* (2007) 63: 1923-1930.

[33] Siyutkin DE, Kucherenko AS, Zlotin SG. A new (*S*)-prolinamide modified by an ionic liquid moiety—a high-performance recoverable catalyst for asymmetric aldol reactions in aqueous media. *Tetrahedron,* (2010) 66: 513-518.

[34] Khan SS, Shah J, Liebscher J. Synthesis of new ionic-liquid-tagged organocatalysts and their application in stereoselective direct aldol reactions *Tetrahedron,* (2010) 66: 5082-5088.

[35] Kochetkov SV, Kucherenko AS, Zlotin SG. (1*R*,2*R*)-Bis[(*S*)-prolinamido]cyclohexane Modified with Ionic Groups: The First C_2-Symmetric Immobilized Organocatalyst for Asymmetric Aldol Reactions in Aqueous Media. *Eur. J. Org. Chem.* (2011) 30: 6128-6133.

[36] Swatloski RP, Holbrey JD, Rogers RD. Ionic liquids are not always green: hydrolysis of 1-butyl-3-methylimidazolium hexafluorophosphate. *Green Chem.*, (2003) 5: 361−363.

[37] Gauchot V, Schmitzer AR. Asymmetric Aldol Reaction Catalyzed by the Anion of an Ionic Liquid. *J. Org. Chem.* (2012) (77) 33: 4917-4923.

[38] Zhang X, Zhao W, Qu C, Yang L, Cui Y. Efficient asymmetric aldol reaction catalyzed by polyvinylidene chloride-supported ionic liquid/L-proline catalyst system. *Tetrahedron: Asymmetry* (2012) (23) 6-7: 468-471.

[39] Yadav GD, Singh S. Direct asymmetric aldol reactions catalyzed by *trans*-4-hydroxy-(*S*)-prolinamide in solvent-free conditions. *Tetrahedron: Asymmetry* (2015) (26) 20: 1156-1166.

[40] Yadav GD, Singh S. (*L*)-Prolinamide imidazolium hexafluorophosphate ionic liquid as an efficient reusable organocatalyst for direct asymmetric aldol reaction in solvent-free condition. *RSC Advance* (2016) 6: 100459-100466.

[41] Deepa, Singh S. Prolinamide and Prolinamide Tagged with Ionic Liquid Intercalated in Bentonite Clay, a Recyclable Catalyst for Asymmetric Aldol Reactions. *ChemistrySelect* (2023) (8) 1: e202202379.

[42] (a) Dalko PL, Moisan L. In the Golden Age of Organocatalysis. *Angew. Chem., Int. Ed.* (2004) (43) 39: 5138-5175; b) Seayad J, List B. Asymmetric organocatalysis. *Org. Biomol. Chem.* (2005) 3: 719-724.

[43] Sulzer-Mosse S, Alexakis A. Ionic Liquid-Supported (ILs) (*S*)-Pyrrolidine Sulfonamide, a Recyclable Organocatalyst for the Highly Enantioselective Michael Addition to Nitroolefins. *Chem. Comm.* (2007) 30: 3123-3135; b) Jarvo ER, Miller SJ. Amino acids and peptides as asymmetric organocatalysts. *Tetrahedron* (2002) 58: 2481-2495.

[44] Krause N, Hoffmann-Roder A. Recent advances in catalytic enantioselective Michael additions. *Synthesis* (2001) 0171-0196; b) Berner OM, Tedeschi L, Enders D. Asymmetric Michael Additions to Nitroalkenes. *Eur. J. Org. Chem.* (2002) 1877-1894; c) Sibi MP, Manyem S. Enantioselective Conjugate Additions. *Tetrahedron* (2000) 56: 8033-8060; d) Christoffers J, Baro A. Construction of Quaternary Stereocenters: New Perspectives through Enantioselective Michael Reactions. *Angew. Chem., Int. Ed.* (2003) 42: 1688-1690.

[45] (a) Hanessian S, Pham V. Catalytic Asymmetric Conjugate Addition of Nitroalkanes to Cycloalkenones. *Org. Lett.* (2000) 2: 2975-2978; b) List B, Pojarliev P, Martin HJ. Efficient Proline-Catalyzed Michael Additions of Unmodified Ketones to Nitro Olefins. *Org. Lett.* (2001) 3: 2423-2425. (c) Enders D, Seki A. Proline-Catalyzed Enantioselective Michael Additions of Ketones to Nitrostyrene. *Synlett*, (2002) 0026-0028; d) Chi Y, Gellman SH. Diphenylprolinol methyl ether: a highly enantioselective catalyst for Michael addition of aldehydes to simple enones. *Org. Lett.* (2005) 7: 4253-4256.

[46] Ishii T, Fujioka S, Sekiguchi Y, Kotsuki H. A New Class of Chiral Pyrrolidine−Pyridine Conjugate Base Catalysts for Use in Asymmetric Michael Addition Reactions. *J. Am. Chem., Soc.* (2004) 126: 9558-9559; b) Mase N, Thayumanavan R, Tanaka F, Barbas CF. III Direct Asymmetric Organocatalytic Michael Reactions of α,α-Disubstituted Aldehydes with β-Nitrostyrenes for the Synthesis of Quaternary Carbon-Containing Products. *Org. Lett.* (2004) 6: 2527-2530; c) Betancort JM, Sakthivel K, Thayumanavan R, Tanaka F, Barbas CF. III *Synthesis* (2004) 1509. (d) Alexakis A, Andrey O. Diamine-Catalyzed Asymmetric Michael Additions of Aldehydes and Ketones to Nitrostyrene. *Org. Lett.* (2002) 4, 3611-3614.

[47] Pansare SV, Pandya K. Simple Diamine- and Triamine-Protonic Acid Catalysts for the Enantioselective Michael Addition of Cyclic Ketones to Nitroalkenes. *J. Am. Chem. Soc.* (2006) 128: 9624-9625; b) Mase N, Watanabe K, Yoda H, Takabe K, Tanaka F, Barbas CF, III Organocatalytic Direct Michael Reaction of Ketones and Aldehydes with β-Nitrostyrene in Brine. *J. Am. Chem. Soc.* (2006) 128: 4966-4967.

[48] Ni B, Zhang Q, Headley AD. Functionalized chiral ionic liquid as recyclable organocatalyst for asymmetric Michael addition to nitrostyrenes. *Green Chem.* (2007) 9: 737-739; b) Zhang Q, Ni B, Headley AD. Asymmetric Michael addition reactions of aldehydes with nitrostyrenes catalyzed by functionalized chiral ionic liquids. *Tetrahedron* (2008) 64: 5091-5097; c) Ni B, Zhang Q, Headley AD. Pyrrolidine-based chiral pyridinium ionic liquids (ILs) as recyclable and highly efficient organocatalysts for the asymmetric Michael addition reactions.

Tetrahedron Lett. (2008) 49: 1249-1252; d) Wu L-Y, Yan Z-Y, Xie Y-X, Niu Y-N, Liang Y-M. Ionic-liquid-supported organocatalyst for the enantioselective Michael addition of ketones to nitroolefins. *Tetrahedron: Asymmetry* (2007) 18: 2086-2090.

[49] Rasalkar MS, Potdar MK, Mohile SS, Salunkhe MM. An ionic liquid influenced L-proline catalysed asymmetric Michael addition of ketones to nitrostyrene. *Journal of Molecular Catalysis A: Chemical*. (2005) 235: 267–270.

[50] Ni B, Zhang Q, Dhungana K, Headley AD. Ionic Liquid-Supported (ILs) (S)-Pyrrolidine Sulfonamide, a Recyclable Organocatalyst for the Highly Enantioselective Michael Addition to Nitroolefins Org. Lett. (2009) 11: 1037-1040.

[51] Luo S, Zhang L, Mi X, Qiao Y, Cheng J-P. Functionalized Chiral Ionic Liquid Catalyzed Enantioselective Desymmetrizations of Prochiral Ketones via Asymmetric Michael Addition Reaction. *J. Org. Chem.* (2007) 72: 9350-9352.

[52] Maltsev OV, Kucherenko AS, Zlotin SG. *O*-TMS-α,α-diphenyl-(*S*)-prolinol Modified with an Ionic Liquid Moiety: A Recoverable Organocatalyst for the Asymmetric Michael Reaction between α,β-Enals and Dialkyl Malonates. *Eur. J. Org. Chem.* (2009) 5134– 5137.

[53] Maltsev OV, Kucherenko AS, Beletskaya IP, Tartakovsky VA, Zlotin SG. Chiral Ionic Liquids Bearing *O*-Silylated α, α-Diphenyl (*S*)- or (*R*)-Prolinol Units: Recoverable Organocatalysts for Asymmetric Michael Addition of Nitroalkanes to α,β-Enals. *Eur. J. Org. Chem.* (2010) 2927– 2933.

[54] Diels, O. Alder. K. Syntheses in the hydroaromatic series. *Justus Liebigs Annalen der Chemie* (1928) 460: 98–122.

[55] Cong H, Porco Jr JA. Total Synthesis of (±)-Sorocenol B Employing Nanoparticle Catalysis. *Org. Lett.* (2012) 14: 2516-2519.

[56] Qi C, Cong H, Cahill K J, Müller P, Johnson RP, Porco, JA. Biomimetic Dehydrogenative Diels–Alder Cycloadditions: Total Syntheses of Brosimones A and B. *Angew. Chem. Int. Ed.* (2013) 52: 8345-8348.

[57] Qi C, Xiong Y, Lux VE, Cong H, Porco JA. Asymmetric Syntheses of the Flavonoid Diels–Alder Natural Products Sanggenons C and O. *J. Am. Chem. Soc.*, (2016) 138: 798-801.

[58] Nomura T, Fukai T, Hano Y, Uzawa J. Structure of Sanggenon C, a Natural Hypotensive Diels-Alder Adduct from Chinese Crude Drug "Sang-bái-pí" (Morus Root Bark). *Heterocycles* (1981) 16: 2141-2148.

[59] Korolev A, Mur V, *Dokl. Akad. Nauk. SSSR Chem. Abstr.* (1948) 107: 251.

[60] Walborsky H, Barash L, Davis T, Inouye Y, Sawada S, Ohno M, Masamune T, Murai A, Orito K, Ono H. Partial asymmetric synthesis in the diels-alder reaction. *Tetrahedron* (1963) 19: 2333-2351.

[61] Kagan HB, Riant O. Catalytic asymmetric Diels Alder reactions. *Chem. Rev.* (1992) 92: 1007-1019.

[62] Ahrendt KA, Borths CJ, MacMillan DWC. New Strategies for Organic Catalysis: The First Highly Enantioselective Organocatalytic Diels−Alder Reaction. *J. Am. Chem. Soc.* (2000) 122: 4243-4244.

[63] Park JK, Sreekanth P, Kim BM. Recycling Chiral Imidazolidin-4-one Catalyst for Asymmetric Diels–Alder Reactions: Screening of Various Ionic Liquids. *Adv. Synth. Catal.* (2004) 346: 49-52.

[64] Nino AD, Bortolini O, Maiuolo L, Garofalo A, Russo B, Sindona G. A sustainable procedure for highly enantioselective organocatalyzed Diels–Alder cycloadditions in homogeneous ionic liquid/water phase. *Tetrahedron Lett.* (2011) 52: 1415-1417.

[65] Shen Z-L, Cheong H-L, Lai Y-C, Loo W-Y, Loh T-P. Application of recyclable ionic liquid-supported imidazolidinonecatalyst in enantioselective Diels–Alder reactions. *Green Chem.* (2012) 14: 2626-2630.

[66] Nino AD, Maiuolo L, Merino P, Nardi M, Procopio A, Lopez DR, Russo B, Algieri V. Efficient Organocatalyst Supported on a Simple Ionic Liquid as a Recoverable System for the Asymmetric Diels–Alder Reaction in the Presence of Water. *ChemCatChem.* (2015)7: 830-835.

[67] Chauhan MS, Kumar P, Singh S. Synthesis of MacMillan catalyst modified with ionic liquid as a recoverable catalyst for asymmetric Diels–Alder reaction. *RSC Adv.* (2015) 5: 52636-52641.

[68] Deepa, Yadav GD, Chaudhary P, Aalam MJ, Meena DR, Singh S. Chiral Imidazolidin-4-one with catalytic amount of Dicationic ionic liquid act as a recoverable and reusable Organocatalyst for asymmetric Diels-Alder reaction. *Chirality* (2020) 32: 64-72.

[69] Singh VK. Practical and Useful Methods for the Enantioselective Reduction of Unsymmetrical Ketones. *Synthesis* (1992) 7: 605-617.

[70] Corey EJ, Helal CJ. Reduction of Carbonyl Compounds with Chiral Oxazaborolidine Catalysts: A New Paradigm for Enantioselective Catalysis and a Powerful New Synthetic Method. *Angew. Chem. Int. Ed.* (1998) 37: 1986-2012.

[71] Corey EJ, Bakshi RK, Shibata S, Chung PC, Singh VK. A stable and easily prepared catalyst for the enantioselective reduction of ketones. Applications to multistep syntheses. *J. Am. Chem. Soc.* (1987) 109: 7925-7926.

[72] Corey EJ, Bakshi RK. A new system for catalytic enantioselective reduction of achiral ketones to chiral alcohols. Synthesis of chiral α-hydroxy acids. *Tetrahedron Lett.* (1990) 31: 611-614.

[73] Brunel JM, Maffei M, Buono G. Enantioselective reduction of ketones with borane, catalyzed by (*S*)-(−)-proline or (*S*)-(+)-prolinol. *Tetrahedron: Asymmetry* (1993) 10: 2255-2260.

[74] Prasad KRK, Joshi NN. An optimized *in situ* procedure for the oxazaborolidine catalyzed enantioselective reduction of prochiral ketones. *Tetrahedron: Asymmetry* (1996) 7: 3147-3152.

[75] Xu J, Wei T, Zhang Q. Effect of Temperature on the Enantioselectivity in the Oxazaborolidine-Catalyzed Asymmetric Reduction of Ketones. Noncatalytic Borane Reduction, a Nonneglectable Factor in the Reduction System. *J. Org. Chem.* (2003) 68: 10146-10146.

[76] Hoogenraad M, Klaus GM, Elders N, Hooijschuur SM, McKay B, Smith AA, Damen EWP. Oxazaborolidine mediated asymmetric ketone reduction: prediction of enantiomeric excess based on catalyst structure. *Tetrahedron: Asymmetry* (2004)15: 519-523.

[77] Xu J, Wei T, Lin S-S Zhang Q. Rationale on the Abnormal Effect of Temperature on the Enantioselectivity in the Asymmetric Borane Reduction of Ketones Catalyzed by L-Prolinol. *Helv. Chim. Acta*. (2005) 88: 180-186.

[78] Wang X, Du J, Liu H, Du D-M, Xu J. Effect of borane source on the enantioselectivity in the enantiopure oxazaborolidine-catalyzed asymmetric borane reduction of ketones. *Heteroat. Chem.* (2007) 18: 740-746.

[79] Zhou Y, Wang WH, Dou W, Tang XL, Liu WS. Synthesis of a new C_2-symmetric chiral catalyst and its application in the catalytic asymmetric borane reduction of prochiral ketones *Chirality* (2008) 20: 110-114.

[80] Dalicsek Z, Pollreisz F, Gömöry A, Soofs T. Recoverable Fluorous CBS Methodology for Asymmetric Reduction of Ketones. *Org. Lett.* (2005) 7: 3243-3246.

[81] Yang S-D, Shi Y, Sun Z-H, Zhao Y-B, Liang Y-M. Asymmetric borane reduction of prochiral ketones using imidazolium-tagged sulfonamide catalyst. *Tetrahedron: Asymmetry* (2006) 17: 1895-1900.

[82] Niu Y-N, Yan Z-Y, Li G-Q, Wei H-L, Gao G-L, Wu L-Y, Liang Y-M. 1,2,3-Triazole-linked dendrimers as a support for functionalized and recoverable catalysts for asymmetric borane reduction of prochiral ketones. *Tetrahedron: Asymmetry* (2008) 19: 912-920.

[83] Du D-M, Fang T, Xu J, Zhang S-W. Structurally Well-Defined, Recoverable C_3-Symmetric Tris(β-hydroxy phosphoramide)-Catalyzed Enantioselective Borane Reduction of Ketones. *Org. Lett.* (2006) 8: 1327-1330.

[84] Price MD, Sui JK, Kurth MJ, Schore NE. Oxazaborolidines as Functional Monomers: Ketone Reduction Using Polymer-Supported Corey, Bakshi, and Shibata Catalysts. *J. Org. Chem.* (2002) 67: 8086-8089.

[85] Kell RJ, Hodge P, Snedden P, Watson D. Towards more chemically robust polymer-supported chiral catalysts: α, α–diphenyl-L-prolinol based catalysts for the reduction of prochiral ketones with borane. *Org. Biomol. Chem.* (2003) 1: 3238-3243.

[86] Degni S, Wilem C-E, Rosling A. Highly catalytic enantioselective reduction of aromatic ketones using chiral polymer-supported Corey, Bakshi, and Shibata catalysts. *Tetrahedron: Asymmetry* (2004) 15: 1495-1499.

[87] Singh A, Chopra HK. New benzimidazolium-based chiral ionic liquids: synthesis and application in enantioselective sodium borohydride reductions in water. *Tetrahedron: Asymmetry* (2016) 27: 448–453.

[88] Chauhan MS, Singh S. Asymmetric reduction of ketones catalyzed by α,α-diphenyl-(L)-prolinol modified with imidazolium ionic liquid and $BH_3{\cdot}SMe_2$ as a recoverable catalyst. *J. Mol. Catal. A: Chem.* (2015) 398: 184–189.

[89] Olah GA, Khirsnamurti R, Prakash GKS. Comprehensive Organic synthesis (Ed.: B. M. Trost, I. Fleming), Pergamon, Oxford (1991) 3: 293; b) Roberts, R. M., Khalaf, A. A. (1984) Friedel-Crafts Alkylation Chemistry Marcel Dekker, New York; c) Olah, G. A. (1973) (Ed.), FriedelCrafts Chemistry Wiley New York; d) Olah G. A. (1963) (Ed.), Friedel-Crafts and Related Reaction Wiley-Interscience, New York, 65: 1.

[90] (a) List B, Lerner, RA, Barbas CF. Proline-Catalyzed Direct Asymmetric Aldol Reactions. *J. Am. Chem. Soc.,* (2000) 122: 2395-2396; b) Northup AB, MacMillan DWC. The First General Enantioselective Catalytic Diels−Alder Reaction with Simple α,β-Unsaturated Ketones. *J. Am. Chem. Soc.,* (2002) 124: 2458-2460.

[91] Paras NA, MacMillan DWC. New Strategies in Organic Catalysis: The First Enantioselective Organocatalytic Friedel−Crafts Alkylation. *J. Am. Chem. Soc.,* (2001) 123: 4370-4371.

[92] Austin JF, MacMillan DWC. Enantioselective Organocatalytic Indole Alkylations. Design of a New and Highly Effective Chiral Amine for Iminium Catalysis. *J. Am. Chem. Soc.*, (2002) 124: 1172-1173.

[93] Singh GS, Desta ZY. Isatins As Privileged Molecules in Design and Synthesis of Spiro-Fused Cyclic Frameworks. *Chem. Rev.,* (2012) 112: 6104-6155.

[94] Liang X, Fan J, Shi F, Su W. Imidazolethiones: novel and efficient organocatalysts for asymmetric Friedel–Crafts alkylation. *Tetrahedron Lett.,* (2010) 51: 2505-2507.

[95] Hack D, Enders D. Asymmetric Organocatalytic Michael Addition of Pyrroles to Enones by Cinchona Alkaloid-Derived Primary Amines. *Synthesis*, (2013) 24: 2904-2912.

[96] Qiao S, Mo J, Wilcox CB, Jiang B, Li G. Chiral GAP catalysts of phosphonylated imidazolidinones and their applications in asymmetric Diels–Alder and Friedel–Crafts reactions. *Org. Biomol. Chem.,* (2017) 15: 1718-1724.

[97] Bonini BF, Capito E, Franchini MC, Fochi M, Ricci A, Zwanenburg B. Aziridin-2-yl methanols as organocatalysts in Diels–Alder reactions and Friedel–Crafts alkylations of *N*-methyl-pyrrole and *N*-methyl-indole. *Tetrahedron: Asymmetry*, (2006) 17: 3135-3143.

[98] Dündar E, Tanyeli C. Enantioselective Friedel-Crafts alkylation of indole with nitroalkenes in the presence of bifunctional squaramide organocatalysts. *Tetrahedron Lett.* (2021) 73: 153153.

[99] Shibatomi K, Narayama A, Abe Y, Iwasa S. Practical synthesis of 4,4,4-trifluorocrotonaldehyde: a versatile precursor for the enantioselective formation of trifluoromethylated stereogenic centers *via* organocatalytic 1,4-additions. *Chem. Commun.*, (2012) 48: 7380.

[100] (a) Terrason V, Figueiredo RM, Campagne J. M. Organocatalyzed Asymmetric Friedel–Crafts Reactions. (2010) *Eur. J. Chem.*, 2635-2655; b) Deepa, Singh S. Recent Development of Recoverable MacMillan Catalyst in Asymmetric Organic Transformations. *Adv. Synth. Catal.*, (2021) 363: 629–656.

[101] Aalam MJ, Deepa, Singh S. Enantioselective Friedel-Crafts Alkylation of Indoles with *α, β*-Unsaturated Aldehydes Catalyzed by Recyclable Second-generation MacMillan Catalysts. *Eur. J. Org. Chem.,* (2022) e202200978.

[102] Deepa, Aalam MJ, Kumar P, Singh S. Enantioselective Friedel-Crafts reaction between indoles and *α,β*-unsaturated aldehydes catalyzed by recyclable *α,α*-diarylprolinol derived chiral ionic liquids. *Tetrahedron Letters,* (2023) 116: 154343.

[103] (a) Tron GC., Minassi, A., Appendino, G. Pietro Biginelli: The Man Behind the Reaction. (2011) *Eur. J. Org. Chem.*, 5541-5550; b) Biginelli P. *Gazz. Chim. Ital.*, (1893) 23: 360.

[104] Shkurko OP, Tolstikova TG, Sedova VF. Dihydropyrimidin-2(1*H*)-one and its analogues as a platform for the design and synthesis of new biologically active compounds. *Russ. Chem. Rev.*, (2016) 85: 1056.

[105] Maliga Z, Kappor TM, Mitchison T. Evidence that Monastrol Is an Allosteric Inhibitor of the Mitotic Kinesin Eg5. *J. Chem. Biol.*, (2002) 9: 989-996.

[106] (a) Ashok M, Holla BS, Kumari NS. Convenient one pot synthesis of some novel derivatives of thiazolo[2,3-*b*]dihydropyrimidinone possessing 4-methylthiophenyl moiety and evaluation of their antibacterial and antifungal activities. *Eur. J. Med. Chem.*, (2007) 42: 380-385; b) Ghorab MM, Mohamed YA, Mohamed SA, Ammar Y. A. *Phosphorus Sulfur Silicon Relat. Elem.*, (1996) 108: 251.

[107] Bozsing D, Sohar P, Gigler G, Kovacs G. Synthesis and pharmacological study of new 3,4-dihydro-2*H*,6*H*-pyrimido-[2,1-*b*][1,3]thiazines. *Eur. J. Med. Chem.*, (1996) 31: 663-668.

[108] Deepa, Aalam MJ, Singh S. Enantioselective Biginelli Reaction Catalyzed by (*L*)-Prolinamide Containing Imidazolium Ionic Liquid. *ChemistrySelect*, (2022) 7: e202103918.

Biographical Sketches

Prof. Surendra Singh: Dr. Surendra Singh was born in 1978 in Rajasthan (India). He received his MSc. in Chemistry in 2000. He has completed his PhD. (2006) in the field of asymmetric catalysis at the Central Salt and Marine Chemical Research Institute, Bhavnagar (India), where he worked on development of chiral catalysts for asymmetric organic transformations under the supervision of late Dr. (Mrs.) R. I. Kureshy. He spent three years at the University College Dublin, Ireland as apostdoctoral researcher in Prof. Pat Guiry's group. Currently, he is a Professor at the Department of Chemistry, University of Delhi. Prof. Singh has published 13 publications in international journal of repute in last three years.

Dr. Geeta Devi Yadav: Dr. Geeta Devi Yadav was born in 1989 in Rajasthan (India). She received her MSc. in Chemistry in 2011. She completed her PhD. in the field of asymmetric catalysis at the University of Delhi, Delhi (India) under the supervision of Dr. Surendra Singh in 2017. Her research interests include the development of organocatalysts in asymmetric organic transformations. Currently, she is working as Women Scientist in Department of Chemistry, University of Delhi, Delhi (India).

Dr. Priyanka Jhajharia: Dr. Priyanka Jhajharia was born in 1981 in Rajasthan (india). She received her MSc. in Chemistry in 2005. She has completed her PhD (2011) in the field of Organophophates at Department of Chemistry, University of Rajasthan under the supervision of Prof. Gita Seth. Currently, She is Associate Professor in Department of Chemistry, Kirori Mal College, University of Delhi, Delhi (India).

Chapter 3

Design of Imidazolium and Pyridinium-Based Ionic Liquid Ligands as Fluorescent Biosensors for G-Quadruplex Detection

Alisson V. Paz, MS
Victor S. Pereira, MS
Clarissa P. Frizzo*, PhD
and Caroline R. Bender†, PhD
Department of Chemistry, Federal University of Santa Maria, Santa Maria, RS, Brazil

Abstract

The fluorescent chemosensor is a class of compounds whose fluorescence emission alternates when an interaction occurs with a specific analyte. Over many decades, various analytes (e.g., metal ions, pH indicators, amino acids, and inorganic anions) have been targeted using fluorescent chemosensors. Furthermore, many organic molecules (e.g., biomolecules) have also been the target of research involving the use of fluorescent chemosensors to better understand biological processes. Among various fluorescent chemosensor structures, heterocyclic moieties containing nitrogen or oxygen are often used, due to the heteroatom lone pairs participating in the resonance of π-conjugated systems, which can manifest fluorescence emission. Heterocyclic structures also have great potential for biocompatibility, due to their similarity to biomolecules, which contain heterocyclic moieties like vitamins, antibodies, and alkaloids. Heterocyclic structures

* Corresponding Author's Email: clarissa.frizzo@gmail.com.
† Corresponding Author's Email: carolinerbender@gmail.com.

In: Imidazolium
Editor: Stephen A. Reyes
ISBN: 979-8-89113-425-6

containing nitrogen can undergo quaternization reactions to form ionic liquids. The formation of ionic species can enhance many properties of the sensor; for example, solubility, electrostatic interactions, hydrogen bonding, or even fluorescence emission. Imidazolium and pyridinium are the most common cationic moieties present in fluorescent chemosensors based on heterocyclic structures. These two cationic species are well known in the field of organic salts and ionic liquids, and knowledge of the properties of this class of compounds has been applied to fluorescent sensors. Among the biomolecules targeted by sensors, G-quadruplexes (G4s) are guanine rich structures found in DNA and RNA, related to the replication and maintenance of genomic stability. The detection of G4s is a field of investigation that has emerged in recent decades, due to G4 structures showing a relationship to replication, genomic stability, and many other biological events important in the treatment and prevention of diseases such as cancer. Thus, the development of fluorescent chemosensors to target these structures became crucial for elucidating their role in genetic processes and maintaining human health.

Keywords: G-quadruplex, biosensors, fluorescence, ionic liquids

Abbreviations and Symbols

ATP	adenosine triphosphate
BMVC	3,6-bis(1-methyl-4-vinylpyridinium)carbazolediiodide
DNA	deoxyribonucleic acid
FPAS	fused polycyclic aromatic systems
FRET	Förster resonance energy transfer
GTP	guanosine triphosphate
G4	g-quadruplex
ICT	intramolecular charge transfer
IL	ionic liquid
IMT	N-Isopropyl-2-(4-N,N-dimethylanilino)-6-methylbenzothiazole
mtDNA	mitochondrial DNA
NFAS	non-fused aromatic systems
PET	photoinduced electron transfer
RNA	ribonucleic acid
TO	thiazole orange
UV	ultraviolet

Introduction

Over the last decade, studies on the development and discovery of chemical structures for detecting and treating various diseases have increased exponentially. Interest in G-quadruplexes (G4s) has increased due to reports detailing their existence in human cells, in which it has been suggested that the G4s of DNA are closely linked to the replication and maintenance of genomic stability (Arola & Vilar, 2008; Burge et al. 2006). Moreover, it has been proposed that the G4s of DNA are epigenetic factors that regulate the expression of cancer-related genes (Burge et al. 2006; Rhodes & Lipps, 2015; Varshney et al. 2020). G4s, which are one of the most important nonclassical secondary structures of nucleic acids, consist of four guanine molecules stabilized by hydrogen bonds. Some molecules, namely G4 ligands, can connect to G4s and generate cellular responses that are correlated with their functions (Arola & Vilar, 2008; Ida et al. 2019; Ma et al. 2015).

Polyaromatic heterocyclic structures, such as acridines, imidazole, benzimidazole, quinoline, and pyrazole, bind to G4s and generally have strong π-π stacking interactions with DNA G4s (Arola & Vilar, 2008; Chilka et al. 2019; Han et al. 2001; Ma et al. 2015; Yuan et al. 2020). Additionally, charged hydrophilic groups induce interactions with G4s because of the electrostatic interaction of the ligand cation with the phosphate groups, which attracts the sensor to the G4 connected to ribose (RNA) or deoxyribose (DNA), thereby increasing sensor selectivity (Chilka et al. 2019; Han et al. 2001; Lai et al. 2014; She et al. 2022). In addition to the affinity between G4s and ligands, the sensor specifications depend on the solubility of the ligand in the biological medium and its ability to diffuse across the plasma membrane. In this context, charged polyaromatic heterocyclic structures, or the insertion of other charged (and polar) structures in covalent form into the same molecular entity as the polyaromatic heterocyclic structure, may be a rational strategy for discovering a selective and efficient sensor for G4s.

Ionic liquids (ILs) are suitable candidates with polar and charged structures. The numerous possible cation–anion combinations in ILs means that the properties of these compounds (e.g., hydrosolubility, liposolubility, and inter- and intramolecular interactions) can be modulated (Lai et al. 2014; Liu et al. 2013; Lu et al. 2016; Souza et al. 2020; Tian et al. 2016). Among the many cations used in IL structures, imidazolium and pyridinium are the cationic moieties used in studies on G4 ligands. The pyridinium moiety and its derivatives have been extensively investigated as sensors for biological analytes (Campbell et al. 2014; Chen et al. 2011; Leduskrasts & Suna, 2019), which is an excellent strategy for detecting biomolecules involved in mitochondrial processes, because the positively charged pyridinium structure is attracted to the negative mitochondrial membrane potential (Guo et al. 2016; She et al. 2022; Xu & Xu, 2016).

For the imidazolium moiety and its derivatives, the relatively high acidity (pKa of 21–23) of the hydrogen atom at position 2 (C_2-H) affects the signal transduction of the recognition processes (Xu et al. 2010). Moreover, the introduction of a positively charged group into the imidazolium ring enhances antiproliferative effects. Additionally, the cationic structure of pyridinium- and imidazolium-based ligands induces intermolecular interactions (π-π and/or π^+- π stacking, and CH- π or H-bonding) with other aromatic structures possessing symmetry similar to the analytes (Campbell et al. 2014; Fujiwara et al. 2016; Singh et al. 2019). Intermolecular interactions and hydrogen bonding between the sensor and the analyte increase or eliminate fluorescence emission via mechanisms such as photoinduced electron transfer (PET) or the quenching of intra- or intermolecular interactions (Daly et al. 2015; Jana et al. 2021; Leduskrasts & Suna, 2019; Sarkar et al. 2011). The insertion of a charged structure into a polyaromatic heterocyclic fluorophore and its conversion into an IL combines the properties of both species.

Based on the aforementioned, this chapter provides a review of G4 organization, including the structures and mechanisms for G4 detection. Additionally, this chapter highlights the advantages of G4 ligand structures based on imidazolium and pyridinium ILs, by combining them with a fluorescent heterocyclic nucleus, thus demonstrating the influence that the charge has on ligand properties. This chapter also contributes to the systematization of knowledge, and provides insights into future prospects in the field. A search of the Web of Science database — using the terms "biosensor," "G-quadruplex," and "ionic liquid" — yielded 415 articles published over the last decade. Adding the terms "imidazolium-based" and "pyridinium-based" yielded 19 and 14 articles, respectively, indicating that this is an expansive field to be explored.

G-Quadruplex

Deoxyribonucleic acid (DNA) consists of two helical chains that are coiled along an axis and form a rotational double helix. It is a long polymer made up basically of nucleotides, whose main chain consists of sugar and phosphate molecules joined by phosphodiester bonds (Sen & Gilbert, 1988). The most known and widely studied conformational form involves four nitrogenous bases — adenine, thymine, cytosine, or guanine — bound to the sugar molecule stabilized by hydrogen bonds. However, DNA can take on other conformations related to certain nucleotide sequences. One of the conformations that can be adopted is the G4, which arises from guanine-rich sequences and can play an extremely important role in a multitude of in vivo processes (Ou et al. 2008; D. Sun et al. 1997; J. Xu et al. 2021) (Figure 1).

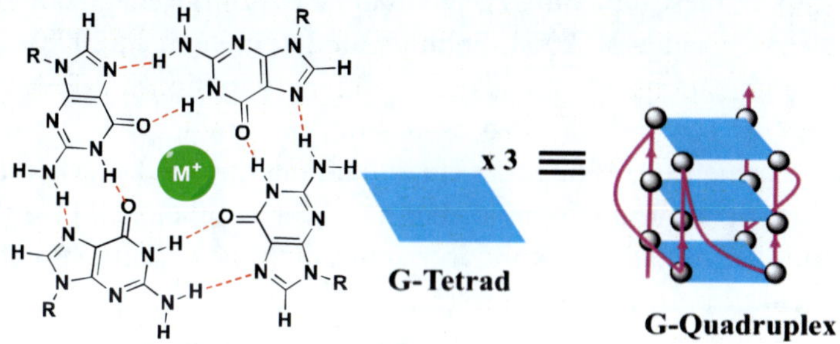

Figure 1. Formation of the G-Quadruplex.

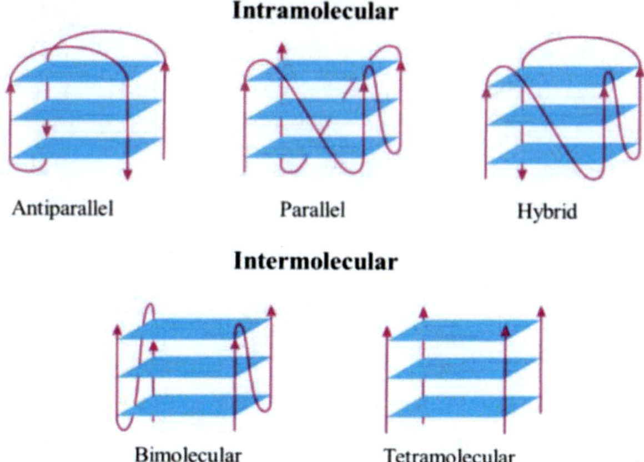

Figure 2. Schematic representation of the different G4 topologies and classifications.

G4s have been widely studied due to their presence in telomeric ends and in other biologically relevant regions of the human genome (Bates et al. 2007). Generally, G4s are secondary nucleic acid structures that are formed within higher-order DNA and RNA structures. These guanine-rich sequences enable the modulation of gene transcription and they are organized by a core and at least two stacked tetrads (known as a G-tetrad, which consists of four guanines aligned in the same plane) that are stabilized by hydrogen bonds (Burge et al. 2006) — see Figure 1. The G-tetrads stack on top of each other through π-π bonds to form the G4s (Yuan et al. 2020).

G4s can be formed with one, two, or four guanine-rich strands, and this requires the use of a medium containing monovalent cations (Na^+ and K^+) and/or small ligands to be present in the biological environment to interact electrostatically with carbonyl oxygen atoms for the formation and stabilization of these structures (Bates et al. 2007; Burge et al. 2006; Z. Y. Sun et al. 2019; de Cian et al. 2008; Folini et al. 2009; Ou et al. 2008). Given the various combinations of strands, as well as variations in loop size and sequence, G4s can be classified as follows: intramolecular, when they are formed by sequences of the same chain; and intermolecular, when they have two or four chains. The intermolecular ones can assume parallel, antiparallel, and hybrid conformations; while intermolecular G4s assume bimolecular or tetramolecular conformations (see Figure 2).

Interest in G4s has increased following reports of the existence of these structures in human cells (Biffi, di Antonio, et al. 2014; Biffi et al. 2013; Murat & Balasubramanian, 2014) and their supposed role in biological regulation (Rhodes & Lipps, 2015). It is suggested that DNA G4s are closely linked with the replication and maintenance of genomic stability. In particular, biological events that depend on the presence of single-stranded DNA may be influenced by the formation of G4s. It has also been proposed that DNA G4s are epigenetic factors that regulate the expression of cancer-related genes (Murat et al. 2011). From another perspective, G4s could be potential targets for cancer treatment. In normal cells, helicase, which unwinds the DNA double helix for replication, can deal with cancer-related genes. However, in cells with mutant helicases, these structures cannot be unraveled (Neidle et al. 2016). This leads to impediments in the transcription, translation, and replication in specific genes, resulting in cellular malfunction and, ultimately, the development of tumors. Therefore, the development of fluorescent chemosensors capable of selectively identifying and localizing G4s in genetic material is fundamental to understanding their biological role (Ma et al. 2015; Neidle et al. 2016). Additionally, these sensors can be used to study and potentially prevent the formation of tumor cells by stabilizing G4 structures (Lipps and Rhodes, 2009; Murat et al. 2014; Neidle et al. 2016). Figure 3 illustrates how a G4 is formed along one of the DNA strands, in an anti-parallel conformation, in which a sensor would interact with its surface, leading to a fluorescence emission response.

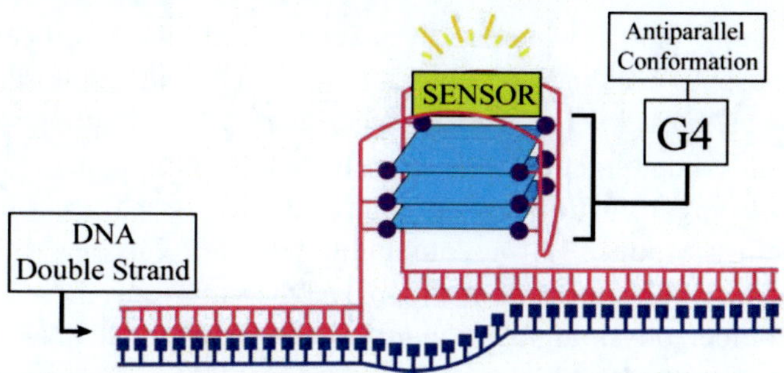

Figure 3. Representation of the interaction of a sensor with DNA G4.

In the human genome, more than 700,000 G4s have been detected in vitro, and their formation is significantly associated with oncogenes (genes linked to the appearance of tumors), tumor suppressors, and somatic copy number alterations related to cancer development (Chambers et al. 2015; Hänsel-Hertsch R., 2017). On the other hand, small molecules called G4 ligands can further stabilize G4s, so that treatment with G4 stabilization ends up disrupting the main biological processes that lead to cell death. Thus, it has been shown that some molecules can bind to G4s and cause cellular responses correlated with their functions (Siddiqui-Jain et al. 2002; Sun et al. 1997).

Overview of the Structures Used as G4 Ligands

With the advances in research on gene regulation via G4s, a multitude of compounds that interact with and/or stabilize G4s to act as chemotherapeutic agents have been studied; for example, the cationic porphyrin (5,10,15,20-tetra(N-methyl-4-pyridyl)porphyrin), which is a G4 stabilizer. This compound has a high affinity for DNA secondary structures present in the telomere region, effectively inhibiting telomerase (Rha et al. 2000). The stabilization of G4s in non-telomeric regions (e.g., proto-oncogenes) provides transcriptional control, resulting in antiproliferative and antitumor activity in various human models, both in vitro and in vivo (Cimino-Reale et al. 2016; Zhang et al. 2015). Stabilization of G4s at telomeres by small molecules inhibits telomerase activity and induces apoptosis in cancer cells (Kim et al. 2002). Selective small G4-stabilizing molecules are linked to the development of new anti-cancer drugs. Thus, destabilization of G4 structures and cancer progression are synergistically connected (Tateishi-Karimata et al. 2018).

The interactions in which small molecules can bind to G4s are: stacking with G4s, groove bonding, loop bonding, and a combination of these various interactions (Arola & Vilar, 2008). According to these binding modes, there are several common structural specificities of G4 ligands, including a polycyclic heteroaromatic core and some charged hydrophilic groups, with the former being responsible for combining with the G4s and the latter for facilitating binding to grooves and loops. At the same time, these ligands must be stable under physiological conditions. G4 ligands fall into three broad classes: (*i*) fused polycyclic aromatic systems (FPAS), (*ii*) macrocycles, and (*iii*) non-fused aromatic systems (NFAS) — see Figure 4. In all cases, fused aromatic polycyclic and macrocyclic systems target the guanines of the outer G-quartets, while the positively charged groups of the side chains interact directly or indirectly, via water molecules, with the grooves or loops.

The electrostatic interaction of the ligand cation with the phosphate groups brings the sensor close to G4, which is connected to ribose or deoxyribose, thus increasing the selectivity of this type of sensor (Han et al. 2001). There are a variety of G4 ligands associated with fluorescence phenomena that allow visualization of secondary structures. The search for an ideal sensor with high G4-binding selectivity and fluorescence enhancement was recently described for several ligands (Chilka et al. 2019).

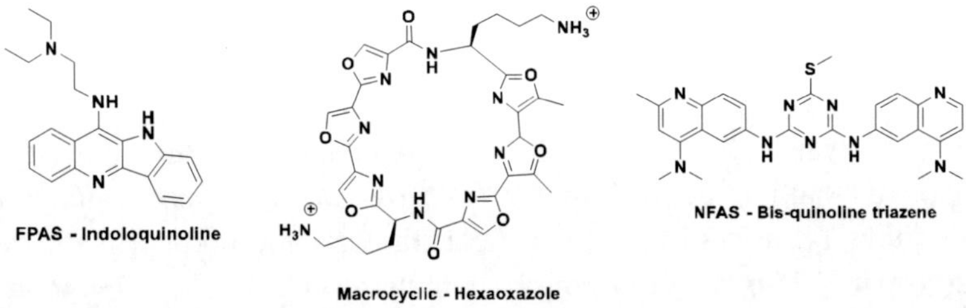

FPAS - Indoloquinoline

Macrocyclic - Hexaoxazole

NFAS - Bis-quinoline triazene

Figure 4. Representative structures of the G4 ligand classes.

The earliest known synthetic dyes (i.e., cyanine derivatives) have been used to visualize G4s. The dye called thiazole orange (TO) is already known as a prominent nucleic acid staining agent and was shown to have a high affinity for G4s. Other fluorescence probes selective for G4s have been built around the pyridinium conjugates' carbazolediiodide derivative (BMVC) and benzothiazole derivative (IMT) portions (Lu et al. 2016). These molecules are potent DNA G4 stabilizing agents and can also be used as sensors to detect DNA G4s (Chilka et al. 2019). BMVC, which is representative of the non-fused aromatic systems (NFAS) classes, has been shown to suppress tumor progression in cancer cells and is also considered a fluorescent sensor for visualizing G4s (Biffi, Tannahill, et al. 2014). Ligands possessing this dual therapeutic and diagnostic property are still limited, since the small molecule must have low cytotoxicity to be qualified as a fluorescent sensor suitable for live cell imaging.

Each of these classes of sensors has advantages and disadvantages. BMVC, for example, is one of the ligands with the greatest selectivity for dyeing and identifying G4s in vivo, but it has several disadvantages, such as low fluorescence intensity, short excitation wavelength, and spectral overlap in the UV region where DNA, proteins, and tissues have very characteristic absorptions. IMT has high fluorescence selectivity for almost all types of DNA and RNA G4 structures (Zhang et al. 2018), which is a great advantage, as many ligands are unsuccessful as sensors because they bind off-target, which substantially decreases the signal-to-noise ratio in live cell analysis. On the other hand, cellular staining of pyridine derivatives is promising when it comes to identifying G4 structures in vivo, as they also show a propensity to bind to other regions of the cell. More research into this aspect is needed to facilitate wider use of these sensors (Chilka et al. 2019b).

Mechanisms for G4 Detection

Organic fluorescent sensors are directly associated with structures with delocalized electrons. Aromatic π-conjugated systems may have the ability to absorb ultraviolet or visible light, but only some of them can emit the absorbed light as a luminescent emission (Czarnik, 1993; Lakowicz, 2006; Valeur & Berberan-Santos, 2011). The occurrence of luminescence as a phenomenon of emission is dependent on the energetic distance between the highest occupied molecular orbital (HOMO) and lowest unoccupied molecular orbital (LUMO), so that the excitation of electrons will occur; for example, in a transition from a π orbital to a π^* orbital (Fu & Finney, 2018; Harbola & Sahni, 1993; Jain et al. 2009). The fluorescence emission phenomenon is a luminescent emission process. The design of a fluorescent sensor usually aims for π-conjugated systems that will display fluorescence emission through a specific mechanism. Some of the most common mechanisms reported are photoinduced electron transfer (PET), intramolecular charge transfer (ICT), and Förster resonance energy transfer (FRET).

The PET process relies on the construction of a species that possesses a fluorophore connected to a receptor (donor). The electron transfer occurs from the donor to the excited state of the fluorophore, and it requires the donor electron's energy to align with the energy gap between the orbitals (e.g., π-π^* and σ-σ^* orbitals of the fluorophore). The donor transfers an electron to the fluorophore in the excited state and blocks the relaxation of the fluorophore's excited electron, which leads to fluorescence emission. Donors are usually electron-rich groups (e.g., heterocyclic compounds containing nitrogen) that can covalently bond to the analyte. When the donor bonds with the analyte, PET no longer occurs, and the fluorescence emission is blocked because the electrons of the donor form a σ-bond with the analyte (Ariztia et al. 2022; Daly et al. 2015; Fu & Finney, 2018; Niu et al. 2023).

Fluorescent sensors usually have different groups attached to the aromatic rings in the π-conjugated system. These groups (electron-donating or electron-withdrawing) have electronic effects on the sensor's molecule. Strategically locating the groups attached to the aromatic system can favor the ICT process, which is based on the charge polarization between a donor site (electron-donating group) and an acceptor site (electron-withdrawing group) — one of the sites may or not be the fluorophore itself. The excited state usually has a stronger dipole moment, where the electron density can migrate from the donor to the acceptor, creating partially negative and positive poles. The change in polarity causes a reorganization in the solvation shell around the sensor to match the charge polarization and lower the system's energy. Two phenomena occur in relation to the solvent shell organization: (1) The solvent shell from the ground state reorganizes to lower the energy of the sensor's excited state, which might occur because the excited state lasts long enough to allow the reorganization of the solvent shell (system's excited state energy is lowered); and (2) The fluorescence emission is a really fast phenomenon, as opposed to the excited state, which does not last long enough for the solvation shell to reorganize immediately to match the ground state again. Subsequently, a ground state molecule with an inadequate solvent shell arises (ground state system's energy is increased). Overall, what happens in the ICT is that the energy level of the excited state decreases and the ground state increases its energy, thus decreasing the energy gap between them, which favors fluorescence emission. ICT usually leads to longer wavelength emission due to the dissipation of the absorbed energy via the exchanges with the solvent shell to rearrange the system's energy (Georgiev et al. 2023; Jana et al. 2021; Jung et al. 2016; Zhang et al. 2018).

It is possible to have more than one emitting moiety in a sensor; for example, two fluorophores connected by a spacer structure. FRET is a mechanism whereby the sensor connects two fluorophores, a donor and an acceptor, and one fluorophore transfers non-radiative energy to the other from the excited state. The FRET process is not a transfer of photon/emission from the donor to the acceptor — it occurs due to a dipole-dipole coupling between donor and acceptor. The FRET process has two general requirements: (1) The fluorescence emission spectrum of the donor and the absorption spectrum of the acceptor must have an overlap, since the energy of this overlap is required for the dipole-dipole coupling to occur; and (2) Due to the importance of the dipole-dipole coupling, the distance between donor and acceptor is important — if the energy levels of the donor and acceptor are too far apart, the coupling will not occur, and the energy transfer is lost. FRET is a very useful tool for protein study, since the distance between donor and acceptor has a direct influence; for example, it is possible to introduce fluorophores in protein structures and map their interaction sites (Hughes et al. 2008; Lakowicz, 2006; Piston & Kremers, 2007; Sekar & Periasamy, 2003; Taya et al. 2018; Wu et al. 2020; Yuan et al. 2013). The FRET process is also used for other biomolecules, such as G4s, and their interactions with other structures (Allain et al. 2006; De Rache & Mergny, 2015; Lee et al. 2019).

G4 Detection Based on Structures with IL Properties

Besides the affinity between G4s and ligands, a useful sensor depends on the following: the solubility in the biological medium, ability to cross the plasma membrane, and low cytotoxicity to the cells. Thus, charged polyaromatic heterocyclic structures or the insertion of other charged (and polar) structures by covalent bonds in the same molecular entity as the polyaromatic heterocyclic structure, could be a rational strategy for finding a fast, selective, and efficient sensor for G4s. Therefore, as polar and charged structures, ILs could be good candidates. The countless possibilities for cation-anion combinations in IL structures means that the properties for this class of compounds (e.g., hydro or liposolubility and ability to perform inter and intramolecular interactions) can be modulated. Among the many cations used in IL structures, imidazolium and pyridinium are the cationic moieties most widely used in the research of G4 ligands.

The use of ILs in the stabilization of DNA and RNA shows that the presence of charges in the molecule can improve aggregation capacity and electronic properties through electrostatic interaction, thus increasing the lifetime of the charge-separated state (Liu et al. 2012; Zhong et al. 1992). Consequently, combining the properties of a fluorophore (a polycyclic aromatic system) with those of a cationic head (e.g., imidazolium or pyridinium) makes it possible to combine the advantages of fluorescent species with those of IL-like species (e.g., low cytotoxicity, solubility, chemical and physiological stability, photostability, and the possibility of modulating the hydrophobicity of the side chain). This strategy for the development of biosensors has been explored by some research groups and has involved a variety of focuses and uses, which will be discussed in the following sections.

Imidazolium-Based ILs as G4 Ligands

Although it is one of the most used cationic moieties in ILs for this specific application, research on the use of imidazolium-based ILs as ligands for G4s is sparse in the scientific literature. It is important to note that the use of imidazole as a detection element in biosensors offers a number of substantial advantages; for example: the remarkable affinity between imidazole and DNA G4 structures; the inherent versatility of this ligand, which means it can be readily functionalized to favor immobilization in sensor substrates such as electrodes or microarray chips; and the ease of detection through the use of a wide range of analytical techniques such as fluorescence spectroscopy, electrochemical techniques, and electron microscopy.

For the imidazolium moiety and its derivatives there is relatively high acidity (pKa of 21–23) in the hydrogen atom bonded to carbon 2 (C_2-H) in water, which can affect the signal transduction of recognition events. Additionally, a positively charged imidazolium ring can enhance antiproliferative effects, due to the formation of covalent bonds with DNA, which leads to damage of the genetic material, thus inhibiting DNA replication (Hu et al. 2021). Imidazolium-based ILs consist of a cationic portion — commonly called the head — linked to a side alkyl chain. A generic imidazolium IL is shown in Figure 5, in which “n” and “m” are the size of different alkyl chains bound to the imidazolium cationic head.

Different fluorescent sensors for the detection of G4s have been reported in the literature; for example, imidazole derivatives (Hu et al. 2015) and ionic species such as benzothiazoles (Wang et al. 2018). Various structural configurations of fluorescent sensors designed for the detection of G4s in order to improve sensitivity and selectivity have been reported. This greater sensitivity becomes imperative due to the distinct conformations exhibited by G4s and the variations in detection that arise from their origin in either DNA or RNA. G4s in DNA and RNA have structural and stability differences, which can be attributed to the additional formation of hydrogen bonds due to the presence of the 2'-OH hydroxyl group in ribose, which confers greater structural stability to guanine dimers (Zaccaria & Fonseca Guerra, 2018).

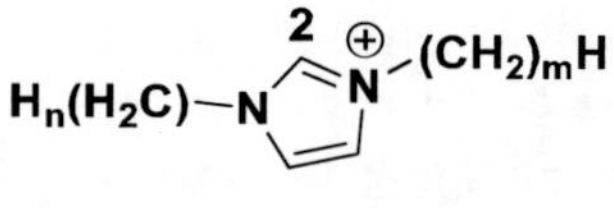

Figure 5. Representation of an imidazolium-based IL.

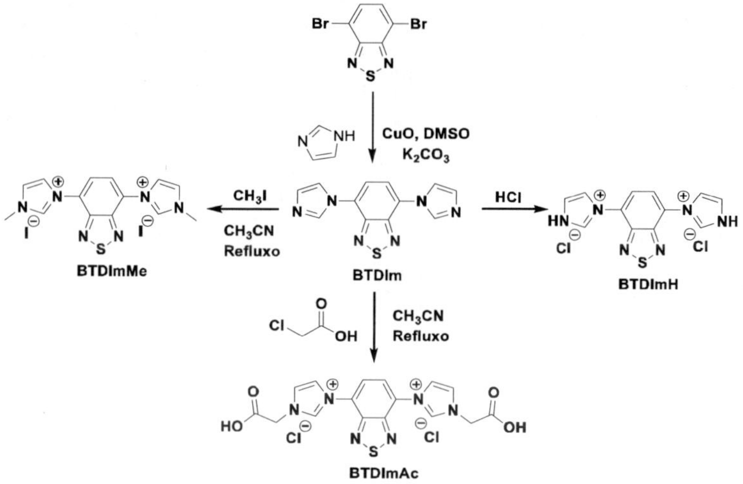

Figure 6. Structures of the compounds synthesized by Souza et al. (2020).

For the development of lysosome-selective dyes, Souza et al. (2020) prepared a series of imidazole-derived ILs linked to the benzothiazole heterocycle. The authors showed that, in addition to solubility in water, the IL conjugated to benzothiazole increased fluorescence intensity and was selective for lysosomes, showing greater photo- and thermo-stability. Chemical structures of the sensors synthesized by Souza et al. (2020) are shown in Figure 6.

Using fluorescence and NMR spectroscopy, Kim et al. (2012) reported a pyrene-imidazolium derivative (Figure 5a) that can recognize G4s with high selectivity. This work is the first example based on the imidazolium derivative, which can detect the G4 directly using the excimer/monomer emission change in the pyrene fluorophore. The biosensor under study was a non-fused aromatic systems (NFAS) type, in which pyrene is the main structure. It was observed that the ratio of excimer and monomer emission of the pyrene fluorophore changes when the sensor binds to guanine sequences on the groove of the G-quartet. This interaction caused an increase in the emission of the excimer in the 186–482 nm spectral range. The affinity constant of the complex was calculated to be 5.5 x 10^5 M^{-1}. In contrast to conventional sensors that recognize the pre-formed G-quartet structure of DNA, this sensor is considered to be a "G-quartet inducer," because it interacts with the guanine-rich region of DNA, thus initiating the process of forming the G-quartet structure. Furthermore, the groove-binding characteristic of 4Gs resulted in only relatively low non-specific toxicity. Xu et al. (2011) reported the synthesis of naphthoimidazolium derivatives and their fluorescence-based recognition of nucleotides in 100% aqueous solutions (Figure 7a and 7c). It should be noted that these structures are part of the fused polycyclic aromatic systems (FPAS) group. The naphthoimidazolium groups in these substances can serve both as a source of hydrogen bonding and additional interactions with the nucleotide bases, as well as a source of fluorescence signaling. Compound 7b, developed by Xu et al. (2011), displays a selective fluorescence enhancement with adenosine triphosphate (ATP) and a selective fluorescence quenching effect with guanosine triphosphate (GTP). Compound 4c does not undergo significant fluorescence changes in the presence of various anions and nucleotides such as ATP, GTP, cytidine triphosphate, thymidine triphosphate, uridine triphosphate, adenosine diphosphate, or adenosine monophosphate.

Recently, Paradis et al. (2022) conducted a comprehensive investigation of the binding interactions of the macrocycle-type sensor TMPyP4 (Figure 8) with the 4G's structure derived from the c-MYC gene. This investigation was performed in both aqueous conditions and imidazolium-based ILs, using a combination of spectroscopic analyses and molecular dynamics simulations. TMPyP4 is widely recognized for its capacity to stabilize 4Gs formed by the c-MYC Pu22 sequence while concurrently diminishing the expression of the oncogenic c-MYC. The UV–Vis and circular dichroism spectroscopies, in conjunction with molecular dynamics simulations, provided a comprehensive elucidation of the intricate binding interactions and associated consequences involving TMPyP4 and four commercial imidazolium-based ILs with distinct side alkyl chain lengths (n = 2, 4, 6, and 8). These investigations were conducted within the context of a 22-mer c-MYC DNA G4 (Pu22). The findings of this study revealed that the four ILs led to a decrease of TMPyP4 binding to Pu22 with the increase of the alkyl side chain length (n = 6 and 8).

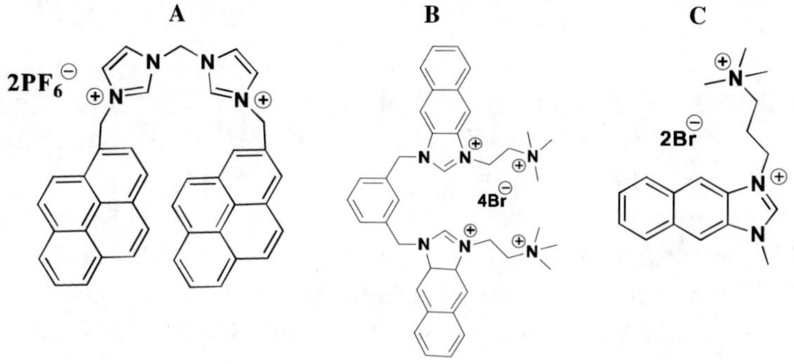

Figure 7. Structures of the compounds synthesized by Xu et al. (2011).

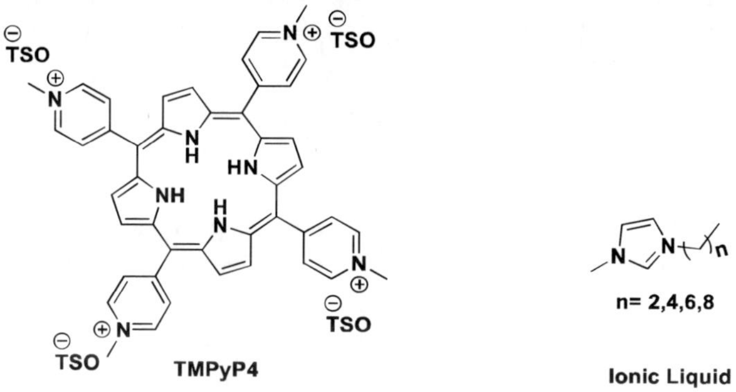

Figure 8. Structures of the compounds evaluated by Paradis et al. (2022).

Molecular dynamics simulations provided understanding of the mechanism by which these ILs bind to TMPyP4 (i.e., by a combination of hydrophobic and electrostatic interactions). While nonpolar IL chains formed hydrophobic interactions with the uppermost G4 layer, the polar ILs' imidazole ring engaged in electrostatic interactions with the DNA backbone. The authors stated that subsequent research should explore the influence of these ILs on various secondary structures within nucleic acids, including their relevance in the context of mRNA vaccines. They also emphasized the limited knowledge regarding the influence that imidazolium-based ILs have on the structural dynamics of G4s and their effect on ligand binding to these complex structures (Paradis et al. 2022).

Pyridinium-Based Ils as G4 Ligands

The pyridinium moiety (Figure 9) and its derivatives are widely used in sensors for biological analytes (Y. Wang et al. 2022; Z. Xu and Xu 2016; Simons and Ikonen 1997). The use of pyridinium-based sensors is well known to be a good strategy for detecting biomolecules present in mitochondrial processes, due to their positively charged structures that are attracted by the mitochondrial negative membrane potential via electrostatic interactions and adequate hydro and liposolubility (Guo et al. 2016; She et al. 2022; Z. Xu and Xu 2016). These properties are very beneficial for the development of pyridinium-based sensors for the detection of G4 structures belonging to mitochondrial DNA (mtDNA) and its regulatory processes (Battogtokh et al. 2018; Huang et al. 2015; Papi et al. 2020; Zhang et al. 2020). The cationic structure of pyridinium-based sensors provides intermolecular interactions (e.g., π-π, π^{+}- π, and CH- π) or hydrogen bonding with other aromatic structures with similar analyte symmetry (Campbell et al. 2014; Chen et al. 2011; Leduskrasts and Suna 2019). Intermolecular interactions and hydrogen bonding between the pyridinium-based sensor and the analyte can increase or extinguish fluorescence emission by PET, quenching intra- or intermolecular interactions and also ICT (inter- or intramolecular charge transfer) due to the pyridinium moiety acting as a donor group with separated anion or in a zwitterionic structure (Hoopes et al. 2022; Leduskrasts and Suna 2019; J. Xu et al. 2017; Reichardt 2008).

Kumari et al. (2019) reported a synthesis of two sensors (see Figure 10) combining the electron-donating chromophore, dimethylamine naphthalene, with an electron-withdrawing group, cationic pyridinium, for the detection of DNA-G4. The authors evaluated the absorption and emission properties of the two sensors in four different solvents: 1,4-dioxane, CH_3CN, CH_3OH, and H_2O. Solvent polarity had a direct influence on absorption: the 1,4-dioxane solution had the most intense absorption band at 465 nm; and the bathochromic absorption shifted as the solvent polarity increased for CH_3CN and CH_3OH. However, H_2O solutions caused a hypsochromic shift of about 45 nm, which the authors attributed to the greater efficiency of H_2O in stabilizing the charged pyridinium moiety, resulting in the absorption in higher energy regions. Compound 2 had the same behavior, with the difference that the absorption shifts between different solutions were more pronounced, due to the stabilization of an extra charged moiety. Polar solvents increase the ICT between the dimethylamine donor group and the pyridinium-withdrawing group, due to stabilization of the polarized system.

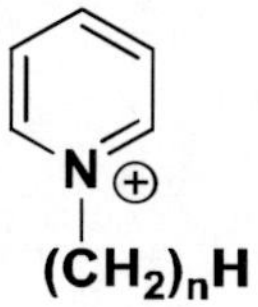

Figure 9. Representation of pyridinium-based ILs.

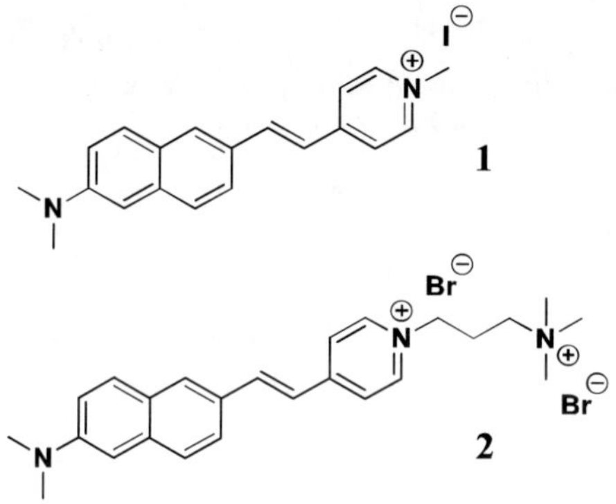

Figure 10. Structures of the compounds evaluated by Kumari et al. 2019.

The fluorescence emission was evaluated through titration with DNA and DNA containing G4 samples. Compound 1 in the presence of DNA had intensity increments of 4.6 units according to the fold-change concept (Mutch et al. 2002) of fluorescence, as well as a hypsochromic shift of 20 nm. In the presence of DNA, compound 2 had intensity increments of 9 units and a hypsochromic shift, indicating a greater affinity with DNA. In the presence of DNA-G4 (antiparallel conformation), compound 1 had an intensity increment of 32 units with a hypsochromic shift; whereas compound 2 had an increment of 73 units. In conclusion, the authors found compound 2 to be a more efficient DNA-G4 sensor, due to the extra cationic moiety that would be interacting strongly with the G4 structure and the phosphate groups, thus highlighting the importance of IL moieties in G4 sensors.

Lai et al. (2014) synthesized and evaluated the DNA-G4 sensor properties of two new symmetric cyanovinyl-pyridinium triphenylamine compounds, referred to by the authors as CPT1 and CPT2 (see Figure 11). The authors used a 22AG DNA sequence (oligonucleotide) as target for the sensors. The data showed that the CPT2 compound had a negligible fluorescence emission when in 10 mM Tris-HCl solution, and a hypochromism (decrease in absorption intensity) of 28% in the presence of 22AG. Nevertheless, there was an increase in the intensity in fluorescence emission of 190 units and a bathochromic shift to a maximum concentration of 10 μM with an emission at 620 nm, in the presence of potassium chloride, to induce the formation of G4 structures in the 22AG solution. There was no significant change in fluorescence emission for compound CPT1 in the presence of 22AG, leading to the conclusion that the monocationic structure does not have a strong interaction with the DNA-G4 structure.

Tests were performed with the CPT2 compound in 22AG with potassium chloride, sodium chloride, or no alkali metal. As expected, the formation of G4 in the presence of K^+ leads to stabler structures — the increase in fluorescence emission was 190 units as opposed to an increase of 90 units with Na^+. Surprisingly, through circular dichroism spectroscopy, it was observed that the CPT2 compound induced the formation of G4 structures in the absence of alkali metal ions. The K^+ ions induced a mixture of the parallel and antiparallel conformations; while in the presence of CPT2, the conformation visualized by the circular dichroism spectra changed from mixed to antiparallel structures, which indicates the selectivity of the sensor for this G4 conformation. The melting point of the 22AG-G4 was also evaluated. An increase of 8°C was observed when compared to the sample in the presence of CPT2, indicating a stabilization of the G4 structure by the CPT2 sensor. Other DNA and RNA sequences were tested by the authors and the majority of them led to the formation of antiparallel G4 structures in the presence of CPT2, thus indicating its role as a fluorescence sensor that can induce and stabilize the formation of G4 structures.

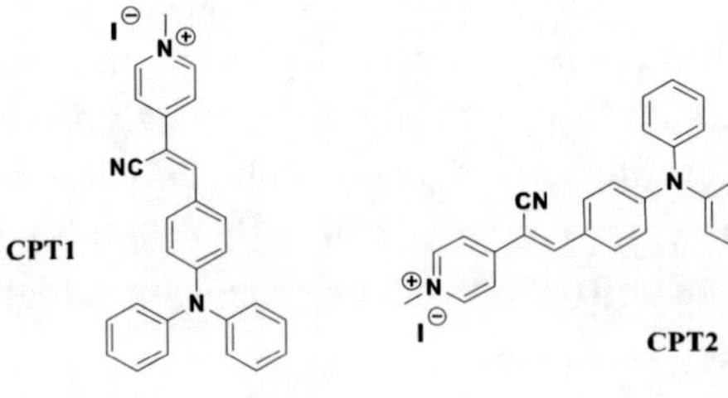

Figure 11. Structures of the compounds evaluated by Lai et al. (2014).

Other structures analogous to pyridinium sensors (e.g., quinolinium) have been reported in the literature; for example, Y. Q. Wang et al. (2018) developed the G4 fluorescent sensor TO-BTZ (see Figure 12) by adding a benzothiazole group to the thiazole orange dye (containing the quinolinium moiety) for detection of G4 structures. The authors observed that, as the original thiazole orange dye, the compound TO-BTZ showed a tendency to form H-aggregates, which are structures that are formed in solution and exhibit a hypsochromic shift when compared to the monomeric structure, as opposed to J-aggregates, which exhibit a bathochromic shift (Eisfeld and Briggs 2006; Gadde et al. 2008; Peyratout, Donath, and Daehne 2001). The aggregation behavior was investigated with Tris-HCl buffer solution and ethanol. The buffer solution, containing TO-BTZ, showed an absorption band at 452 nm; while in ethanol there were two bands (at 483 and 503 nm), thus indicating the presence of aggregate and monomeric forms. Experiments conducted in simulated physiological media with Tris-HCl buffer solution showed weak fluorescence emission at 600 nm for TO-BTZ. With the gradual addition of telomeric G4, with conformation tel24 (hybrid), a new emission peak at 550 nm was increasingly observed. The authors concluded that, due to its specific interactions, the G4-tel24 readily dissipates the aggregates of TO-BTZ to monomers. Other G4 conformations with different nucleic acid sequences were also tested, and the TO-BTZ had weaker responses than the tel24 conformation, thus evidencing its selectivity.

TO-BTZ

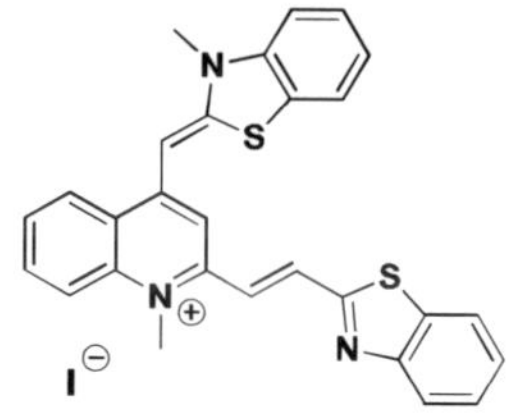

Figure 12. Structures of the compounds evaluated by Kumari et al. (2019).

Other Perspectives

Understanding the distribution of G4s in normal cells versus cancerous cells would promote the development of more selective tools in relation to DNA or RNA G4s at any stage of the cell cycle (Biffi, di Antonio, et al. 2014). Visualization of DNA or RNA G4 structures can be achieved via immunodetection, in situ fluorescence labeling of G4-selective ligands, and small molecular fluorescence sensors. These approaches enable the study of ligand selectivity for DNA or RNA G4s under defined cellular conditions (Biffi et al. 2013). The transient nature of G4 structures and their resolution in the genome makes it challenging to study G4 formation in eukaryotic cells.

Visual tools can help assess the formation of G4s during metabolic processes as well as during cell differentiation (Laguerre et al. 2016).

Identifying the impact, at the molecular level, of G4 structures in the vital processes associated with cancer is likely to open up new avenues in cancer therapy and diagnosis. However, distinguishing the causality of G4 structures in the context of cancer is a challenge. Furthermore, knowledge of the influence that these structures have on cancer initiation and progression is limited, which is due to challenges in detecting and analyzing cellular responses that could be directly correlated to these structures in living cells.

Conclusion

This chapter explored the organization of G4s found in DNA and RNA, highlighted the main fluorescence spectroscopy detection mechanism, and addressed some structures used as G4 ligands. It was shown that numerous compounds of different natures can interact with and/or stabilize G4s, acting as chemotherapeutic agents. It was shown that G4 detection is a promising field of investigation, whose attention from researchers has been increasing in recent decades. This can be attributed to the G4s' link to replication, genomic stability, and many other biological events that are important in the treatment and prevention of diseases.

Additionally, the chapter focused on the advantages of G4 ligand structures derived from imidazolium and pyridinium ILs and combined with a fluorescent nucleus. Although the existing literature in this field is limited, the studies done to date and discussed here have shown promising insights into the potential of these ligands, demonstrating the influence that the nature and charge of these structures have on the G4s' stabilization and selectivity. Nevertheless, further research will be necessary to advance the development of fluorescent chemosensors based on these technologic materials, since the ability to interact selectively with G4 structures can by adjusted by considering the intrinsic versatility of ILs (e.g., by synthetic changes in cations and the size of the alkyl chain and/or anions). This could mean specific property modifications for different applications, which would provide new perspectives for the research and development of biosensors and molecular diagnostic tools.

Thus, the design of imidazolium- and pyridinium-based ILs for selectively stabilizing G4s (in accordance with precise characteristics and conformations) may become decisive in elucidating the possible role that G4s play in genetic processes and maintaining human health. We hope to have contributed regarding perspectives on the use of these ILs as G4 ligands and inspire others about the great number of possibilities to be investigated in this emerging field of research.

Disclaimer

None.

References

Allain C, Monchaud D, and Teulade-Fichou MP. FRET Templated by G-Quadruplex DNA: A Specific Ternary Interaction Using an Original Pair of Donor/Acceptor Partners. *Journal of the American Chemical Society* (2006) 128(36):11890–11893.

Ariztia J, Solmont K, Moïse SN, Specklin P, Heck MP, Lamandé-Langle S, and Kuhnast B. PET/Fluorescence Imaging: An Overview of the Chemical Strategies to Build Dual Imaging Tools. *Bioconjugate Chemistry* (2022) 33(1):24–52.

Arola A, Ramon V. Stabilisation of G-quadruplex DNA by small molecules. *Current Topics in Medicinal Chemistry* (2008) 8(15):1405-1415.

Bates P, Jean-Louis M, and Danzhou Yang. Quartets in G-major: The first international meeting on quadruplex DNA. *EMBO reports* (2007) 8(11):1003-1010.

Battogtokh G, Choi YS, Kang DS, Park SJ, Shim MS, Huh KM, Cho YY, Lee JY, Lee HS, and Kang H C. Mitochondria-Targeting Drug Conjugates for Cytotoxic, Anti-Oxidizing and Sensing Purposes: Current Strategies and Future Perspectives. *Acta Pharmaceutica Sinica B* (2018) 8(6):862-880.

Biffi G, Di Antonio M, Tannahill, D, and Balasubramanian, S. Visualization and selective chemical targeting of RNA G-quadruplex structures in the cytoplasm of human cells. *Nature chemistry* (2014) 6(1):75-80.

Biffi G, Tannahill D, McCafferty J, and Balasubramanian S. Quantitative visualization of DNA G-quadruplex structures in human cells. *Nature chemistry* (2013) 5(3):182-186.

Burge S, Parkinson, GN, Hazel P, Todd AK, and Neidle S. Quadruplex DNA: sequence, topology and structure. *Nucleic acids research* (2006) 34(19):5402-5415.

Campbell P S, Yang M, Pitz D, Cybinska J, and Mudring AV. Highly Luminescent and Color-Tunable Salicylate Ionic Liquids. *Chemistry - A European Journal* (2014) 20(16):4704-4712.

Chambers VS, Marsico G, Boutell JM, Di Antonio M, Smith GP, and Balasubramanian S. High-throughput sequencing of DNA G-quadruplex structures in the human genome. *Nature biotechnology* (2015) 33(8): 877-881.

Chen, W, Elfeky SA, Nonne Y, Male L, Ahmed K, Amiable C, Axe P, Yamada S, James TD, Bull SD, and Fossey J S. A Pyridinium Cation–π Interaction Sensor for the

Fluorescent Detection of Alkyl Halides. *Chemical Communications* (2011) 47(1):253–255

Chilka P, Desai N, and Datta B. Small molecule fluorescent probes for G-quadruplex visualization as potential cancer theranostic agents. *Molecules* (2019) 24(4):752.

De Cian A, Lacroix L, Douarre C, Temime-Smaali N, Trentesaux C, Riou JF, and Mergny JL. Targeting telomeres and telomerase. *Biochimie* (2008) 90(1):131-155.

Cimino-Reale G, Zaffaroni N, and Folini M. Emerging role of G-quadruplex DNA as target in anticancer therapy. *Current Pharmaceutical Design* (2016) 22(44): 6612-6624.

Czarnik AW, and Desvergne JP. *Chemosensors of Ion and Molecule Recognition*. Springer Netherlands. (1997):1-9.

Daly, B, Ling, J and De Silva, AP. Current Developments in Fluorescent PET (Photoinduced Electron Transfer) Sensors and Switches. *Chemical Society Reviews* (2015) 44(13): 4203–4211.

De Rache A, and Mergny JL. Assessment of Selectivity of G-Quadruplex Ligands via an Optimised FRET Melting Assay. *Biochimie* (2015) 115:194–202.

Eisfeld A, and Briggs JS. The J- and H-Bands of Organic Dye Aggregates. *Chemical Physics* (2006) 324(2–3):376–384.

Folini M, Gandellini P, and Zaffaroni N. Targeting the telosome: therapeutic implications. *Biochimica et Biophysica Acta (BBA)-Molecular Basis of Disease* (2009) 1792 (4):309-316.

Fu Y, and Finney NS. Small-Molecule Fluorescent Probes and Their Design. *RSC Advances* (2018) 8:29051-29061.

Fujiwara S, Ichikawa T, and Ohno H. Cation-π Interactions within Aromatic Amino Acid Ionic Liquids: A New Tool for Designing Functional Ionic Liquids. *Journal of Molecular Liquids* (2018) 222:214–217.

Gadde S, Batchelor EK, Weiss JP, Ling Y, and Kaifer AE. Control of H- and J-Aggregate Formation via Host-Guest Complexation Using Cucurbituril Hosts. *Journal of the American Chemical Society* (2008) 130(50):17114–17119.

Georgiev NI, Bakov VV, Anichina KK, and Bojinov VB. Fluorescent Probes as a Tool in Diagnostic and Drug Delivery Systems. *Pharmaceuticals.* (2023) 16(3):381.

Guo L, Zhang R, Sun Y, Tian M, Zhang G, Feng R, Li X, Yu X, and He X. Styrylpyridine Salts-Based Red Emissive Two-Photon Turn-on Probe for Imaging the Plasma Membrane in Living Cells and Tissues. *Analyst* (2016) 141(11):3228–3232.

Han H, Langley DR, Rangan A, and Hurley LH. Selective interactions of cationic porphyrins with G-quadruplex structures. *Journal of the American Chemical Society* (2001) 123(37): 8902-8913.

Hänsel-Hertsch R, Spiegel J, Marsico G, Tannahill D, and Balasubramanian S. Genome-wide mapping of endogenous G-quadruplex DNA structures by chromatin immunoprecipitation and high-throughput sequencing. *Nature protocols* (2018) 13(3):551-564.

Harbola MK, and Sahni V. Theories of Electronic Structure in the Pauli-Correlated Approximation. *Journal of Chemical Education* (1993) 70(11):920–927.

Hoopes CR, Garcia FJ, Sarkar AM, Kuehl NJ, Barkan DT, Collins NL, Meister GE, Bramhall TR, Hsu C, Jones MD, Schirle M, and Taylor MT. Donor-Acceptor Pyridinium Salts for Photo-Induced Electron-Transfer-Driven Modification of

Tryptophan in Peptides, Proteins, and Proteomes Using Visible Light. Journal of the *American Chemical Society* (2022) 144(14):6227–6236.

Hu MH, Chen SB, Guo RJ, Ou TM, Huang ZS, and Tan JH, Development of a highly sensitive fluorescent light-up probe for G-quadruplexes. *Analyst* (2015) 140(13):4616-4625.

Hu Y, Long S, Fu H, She Y, Xu Z, and Yoon J. Revisiting imidazolium receptors for the recognition of anions: Highlighted research during 2010–2019. *Chemical Society Reviews* (2021) 50(1):589-618.

Huang WC, Tseng TY, Chen YT, Chang CC, Wang ZF, Wang CL, Hsu TN, Li P, Chen C, Lin J, Lou P, and Chang T. Direct Evidence of Mitochondrial G-Quadruplex DNA by Using Fluorescent Anti-Cancer Agents. Nucleic Acids Research (2015) 43(21):10102–10113.

Hughes AD, Glenn IC, Patrick AD, Ellington A, and Anslyn EV. A Pattern Recognition Based Fluorescence Quenching Assay for the Detection and Identification of Nitrated Explosive Analytes. *Chemistry - A European Journal* (2008) 14(6):1822–1827.

Ida J, Chan SK, Glökler J, Lim YY, Choong YS, and Lim TS. G-Quadruplexes as an Alternative Recognition Element in Disease-Related Target Sensing. *Molecules.* MDPI AG. (2019) 24(6):1079.

Jain A, Blum C, and Subramaniam V Fluorescence Lifetime Spectroscopy and Imaging of Visible Fluorescent Proteins. In *Advances in Biomedical Engineering*. Elsevier. (2009):147-176.

Jana A, Baruah M, Munan S, and Samanta AS. ICT Based Water-Soluble Fluorescent Probe for Discriminating Mono and Dicarbonyl Species and Analysis in Foods. *Chemical Communications* (2021) 57(52):6380–6383.

Jung HS, Verwilst P, Kim WY, and Kim JS. Fluorescent and Colorimetric Sensors for the Detection of Humidity or Water Content. *Chemical Society Reviews* (2016) 45(5):1242–1256.

Kim MY, Vankayalapati H, Shin-Ya K, Wierzba K, and Hurley LH. Telomestatin, a potent telomerase inhibitor that interacts quite specifically with the human telomeric intramolecular G-quadruplex. *Journal of the American Chemical Society* (2002) 124(10):2098-2099.

Kumari B, Yadav A, Pany SP, Pradeepkumar PI, and Kanvah S. Cationic Red Emitting Fluorophore: A Light up NIR Fluorescent Probe for G4-DNA. *Journal of Photochemistry and Photobiology B: Biology* (2019) 190:128–136.

Laguerre A, Wong JM, and Monchaud D. Direct visualization of both DNA and RNA quadruplexes in human cells via an uncommon spectroscopic method. *Scientific reports* (2016) 6(1):32141.

Lai, H., Y. Xiao, S. Yan, F. Tian, C. Zhong, Y. Liu, X. Weng, and X. Zhou. 2014. Symmetric Cyanovinyl-Pyridinium Triphenylamine: A Novel Fluorescent Switch-on Probe for an Antiparallel G-Quadruplex. *Analyst* (2014) 139(8): 1834–1838.

Lakowicz RJ. *Principles of Fluorescence Spectroscopy* (3rd ed.). Springer US. (2006): 1-23.

Leduskrasts K, and Suna E. Aggregation Induced Emission by Pyridinium-Pyridinium Interactions. *RSC Advances* (2019) 9(1):460–465.

Lee C, McNerney C, and Myong S. *G-Quadruplex and Protein Binding by Single-Molecule FRET Microscopy*. (2019) 2035:309–322.

Lipps HJ, and Rhodes D. G-quadruplex structures: in vivo evidence and function. *Trends in cell biology* (2009) 19(8):414-422.

Liu K, Yao Y, Liu Y, Wang C, Li Z, and Zhang X. Self-assembly of supra-amphiphiles based on dual charge-transfer interactions: from nanosheets to nanofibers. *Langmuir* (2012) 28(29):10697-10702.

Liu H, Zhang L, Chen J, Zhai Y, Zeng Y, and Li L. A Novel Functional Imidazole Fluorescent Ionic Liquid: Simple and Efficient Fluorescent Probes for Superoxide Anion Radicals. *Analytical and Bioanalytical Chemistry* (2013) 405(29):9563–9570.

Lu YJ, Hu D P, Zhang K, Wong WL, and Chow CF. New pyridinium-based fluorescent dyes: A comparison of symmetry and side-group effects on G-Quadruplex DNA binding selectivity and application in live cell imaging. *Biosensors and Bioelectronics* (2016) 81:373-381.

Ma DL, Zhang Z, Wang M, Lu L, Zhong HJ, and Leung CH. Recent developments in G-quadruplex probes. *Chemistry & Biology* (2015) 22(7):812-828.

Murat P, and Balasubramanian S. Existence and consequences of G-quadruplex structures in DNA. *Current opinion in genetics & development* (2014) 25:22-29.

Murat P, Singh Y, and Defrancq E. Methods for investigating G-quadruplex DNA/ligand interactions. *Chemical Society Reviews* (2011) 40(11):5293-5307.

Mutch DM, Berger A, Mansourian R, Rytz A, and Roberts M. The Limit Fold Change Model: A Practical Approach for Selecting Differentially Expressed Genes from Microarray Data. *BMC Bioinformatics* (2002) 3:17.

Neidle, S. Quadruplex nucleic acids as novel therapeutic targets. *Journal of medicinal chemistry* (2016) 59(13):5987-6011.

Niu H, Liu J, O'Connor HM, Gunnlaugsson T, James TD, and Zhang H. Photoinduced Electron Transfer (PeT) Based Fluorescent Probes for Cellular Imaging and Disease Therapy. *Chemical Society Reviews* (2023) 52(7):2322–2357.

Ou TM, Lu YJ, Tan JH, Huang ZS, Wong KY, and Gu LQ. G-quadruplexes: targets in anticancer drug design. *ChemMedChem: Chemistry Enabling Drug Discovery* (2008) 3(5):690-713.

Papi F, Bazzicalupi C, Ferraroni M, Ciolli G, Lombardi P, Khan A Y, Kumar GS, and Gratteri P. Pyridine Derivative of the Natural Alkaloid Berberine as Human Telomeric G4-DNA Binder: A Solution and Solid-State Study. *ACS Medicinal Chemistry Letters* (2020) 11(5):645–650.

Paradis NJ, Clark A, Gogoj H, Lakernick PM, Vaden TD, and Wu C. To probe the binding of TMPyP4 to c-MYC G-quadruplex with in water and in imidazolium-based ionic liquids using spectroscopy coupled with molecular dynamics simulations. *Journal of Molecular Liquids*(2022) 365:120097.

Peyratout C, Donath E, and Daehne L. Electrostatic Interactions of Cationic Dyes with Negatively Charged Polyelectrolytes in Aqueous Solution. *Journal of Photochemistry and Photobiology A: Chemistry* (2001) 142(1):51-57.

Piston DW, and Kremers G. Fluorescent Protein FRET: The Good, the Bad and the Ugly. *Trends in Biochemical Sciences* (2007) 32(9):407–414.

Reichardt C. Pyridinium-N-Phenolate Betaine Dyes as Empirical Indicators of Solvent Polarity: Some New Findings. *Pure and Applied Chemistry* (2008) 80(7):1415–1432.

Rha SY, Izbicka E, Lawrence R, Davidson K, Sun D, Moyer MP, G. Roodman D, Hurley L, and Von Hoff, D. Effect of telomere and telomerase interactive agents on human tumor and normal cell lines. *Clinical Cancer Research* (2000) 6(3):987-993.

Rhodes D, and Lipps HJ. G-quadruplexes and their regulatory roles in biology. *Nucleic acids research* (2015) 43(18):8627-8637.

Sarkar S, Pramanik R, Ghatak C, Rao VG, and Sarkar N. Photoinduced Intermolecular Electron Transfer in a Room Temperature Imidazolium Ionic Liquid: An Excitation Wavelength Dependence Study. *Chemical Physics Letters* (2011) 506(4–6):211–216.

Sekar RB, and Periasamy A. Fluorescence Resonance Energy Transfer (FRET) Microscopy Imaging of Live Cell Protein Localizations. The Journal of Cell Biology (2003) 160(5):629–633.

Sen D, and Gilbert W. Formation of parallel four-stranded complexes by guanine-rich motifs in DNA and its implications for meiosis. *Nature* (1988) 334(6180):364-366.

She M T, Yang J W, Zheng B X, Long W, Huang X H, Luo J R, and Chen Z X. Design Mitochondria-Specific Fluorescent Turn-on Probes Targeting G-Quadruplexes for Live Cell Imaging and Mitophagy Monitoring Study. *Chemical Engineering Journal* (2022) 446(2):136947.

Siddiqui-Jain A, Grand CL, Bearss DJ, and Hurley LH. Direct evidence for a G-quadruplex in a promoter region and its targeting with a small molecule to repress c-MYC transcription. *Proceedings of the National Academy of Sciences* (2002) 99(18):11593-11598.

Simons K, and Ikonen E. Functional Rafts in Cell Membranes. *Nature* (1997) 387 (6633):569–572.

Singh O, Singla P, Aswal VK, and Mahajan RK. Impact of Aromatic Counter-Ions Charge Delocalization on the Micellization Behavior of Surface-Active Ionic Liquids. *Langmuir* (2019) 35(45):14586–14595.

Souza VS, Corrêa JR, Carvalho PH, Zanotto GM, Matiello GI, Guido BC, Gatto CC, Ebeling G, Gonçalves P F B, Dupont J, and Neto B A. Appending ionic liquids to fluorescent benzothiadiazole derivatives: light up and selective lysosome staining. *Sensors and Actuators B: Chemical* (2020) 321:128530.

Sun D, Thompson B, Cathers BE, Salazar M, Kerwin SM, Trent JO, Jenkins TS, Neidle S, and Hurley L H. Inhibition of human telomerase by a G-quadruplex-interactive compound. *Journal of medicinal chemistry* (1997) 40(14):2113-2116.

Sun Z, Wang XN, Cheng SQ, Su XX, and Ou TM. Developing novel G-quadruplex ligands: From interaction with nucleic acids to interfering with nucleic acid–protein interaction. *Molecules* (2019) 24(3):396.

Tateishi-Karimata H, Kawauchi K, and Sugimoto N. Destabilization of DNA G-quadruplexes by chemical environment changes during tumor progression facilitates transcription. *Journal of the American Chemical Society* (2018) 140(2):642-651.

Taya P, Maiti B, Kumar V, De P, and Satapathi S. Design of a Novel FRET Based Fluorescent Chemosensor and Their Application for Highly Sensitive Detection of Nitroaromatics. *Sensors and Actuators, B: Chemical* (2018) 255:2628–2634.

Tian X, Qi X, Liu X, and Zhang Q. Selective Detection of Picric Acid by a Fluorescent Ionic Liquid Chemosensor. *Sensors and Actuators B: Chemical* (2016) 229:520–527.

Valeur B, and Berberan-Santos MN. A Brief History of Fluorescence and Phosphorescence before the Emergence of Quantum Theory. *Journal of Chemical Education* (2011) 88(6):731-738.

Varshney D, Spiegel J, Zyner K, Tannahill D, and Balasubramanian S. The Regulation and Functions of DNA and RNA G-Quadruplexes. Nature Reviews Molecular Cell Biology. *Nature Research (*2020) 21(8):459-474.

Wang YQ, Hu MH, Guo RJ, Chen SB, Huang ZS, and Tan JH. Tuning the selectivity of a commercial cyanine nucleic acid dye for preferential sensing of hybrid telomeric G-quadruplex DNA. *Sensors and Actuators B: Chemical* (2018) 266:187-194.

Wang Y, Chen N, Pan Z, Ye Z, Yuan J, Zeng Y, and Long W. A Smart Small Molecule as Specific Fluorescent Probe for Sensitive Recognition of Mitochondrial DNA G-Quadruplexes. *Chemical Engineering Journal* (2022) 441:135977.

Wu L, Huang C, Emery BP, Sedgwick AC, Bull SD, He X, Tian H, Yoon J, Sessler JL, and James TD. Förster Resonance Energy Transfer (FRET)-Based Small-Molecule Sensors and Imaging Agents. *Chemical Society Reviews* (2020) 49(15):5110–5139.

Xu J, Zhang B, Jansen M, Goerigk L, Wong W W H, and Ritchie C. Highly Fluorescent Pyridinium Betaines for Light Harvesting. *Angewandte Chemie International Edition* (2017) 56(44):13882–13886.

Xu Z, and Xu L. Fluorescent Probes for the Selective Detection of Chemical Species inside Mitochondria. *Chemical Communications* (2016) 52(6):1094–1119.

Xu Z, Kim SK, and Yoon J. Revisit to Imidazolium Receptors for the Recognition of Anions: Highlighted Research during 2006–2009. *Chemical Society Reviews* (2010) 39(5):1457.

Xu J, Jiang R, He H, Ma C, and Tang Z. Recent advances on G-quadruplex for biosensing, bioimaging and cancer therapy. *TrAC Trends in Analytical Chemistry* (2021) 139:116257.

Xu Z, Song NR, Moon JH, Lee JY, and Yoon J. Bis-and tris-naphthoimidazolium derivatives for the fluorescent recognition of ATP and GTP in 100% aqueous solution. *Organic & Biomolecular Chemistry* (2011) 9(24):8340-8345.

Yuan L, Lin W, Zheng K, and Zhu S. FRET-Based Small-Molecule Fluorescent Probes: Rational Design and Bioimaging Applications. *Accounts of Chemical Research* (2013) 46(7):1462–1473.

Yuan JH, Shao W, Chen SB, Huang ZS, and Tan JH. Recent advances in fluorescent probes for G-quadruplex nucleic acids. *Biochemical and Biophysical Research Communications* (2020) 531(1):18-24.

Zaccaria F, and Fonseca Guerra C. RNA versus DNA G-Quadruplex: The Origin of Increased Stability. *Chemistry–A European Journal* (2018) 24(61):16315-16322.

Zhang W, Chen M, Ling Wu Y, Tanaka Y, Juan Ji Y, Lin Zhang S, Wei C H, and Xu Y. Formation and stabilization of the telomeric antiparallel G-quadruplex and inhibition of telomerase by novel benzothioxanthene derivatives with anti-tumor activity. *Scientific reports* (2015) 5(1):13693.

Zhang S, Sun H, Wang L, Liu Y, Chen H, Chen H, Li Q, Guan A, Liu M, and Tang Y. Real-time monitoring of DNA G-quadruplexes in living cells with a small-molecule fluorescent probe. *Nucleic acids research* (2018) 46(15):7522-7532.

Zhang Y, Wang L, Rao Q, Bu Y, Xu T, Zhu X, Zhang J, Tian Y, and Zhou H. Tuning the Hydrophobicity of Pyridinium-Based Probes to Realize the Mitochondria-Targeted Photodynamic Therapy and Mitophagy Tracking. *Sensors and Actuators, B: Chemical* (2020) 321(15):128460.

Zhong CJ, Kwan WSV, and Miller LL. Self-assembly of delocalized pi-stacks in solution. Assessment of structural effects. *Chemistry of materials* (1992) 4(6): 1423-1428.

Chapter 4

Different Properties and Applications of Imidazolium Ionic Liquids: A DFT Study

Madhulata Shukla*, **PhD**
Department of Chemistry, Gram Bharti College Ramgarh,
Veer Kunwar Singh University, Bihar, India

Abstract

Imidazolium ionic liquids offer a wide range of properties and applications, making them versatile materials in various scientific and industrial fields. Density Functional Theory study gained valuable insights into their thermophysical, electrochemical, and solvation properties. The understanding of these properties enables the design and development of novel imidazolium ILs tailored for specific applications, contributing to the advancement of green chemistry, energy storage, and chemical synthesis.

Keywords: ionic liquids, DFT calculation, interactions

1. Introduction

Ionic liquids (ILs) are an important class of materials that have received increasing attention due to their unique properties and potential applications in various fields such as catalysis, electrochemistry, and separation science (Welton 2002). It is worth noting that ILs have attracted significant attention

* Corresponding Author's Email: madhu1.shukla@gmail.com.

In: Imidazolium
Editor: Stephen A. Reyes
ISBN: 979-8-89113-425-6

in various fields due to their exceptional properties. They have been widely explored as green solvents, electrolytes in energy storage devices, and catalysts in chemical reactions (Z. Wang et al. 2020). By carefully designing the structure and composition of ILs, researchers can fine-tune their properties for specific applications. ILs exhibit a unique behavior compared to conventional solvents. In addition to typical ionic and covalent interactions, ILs also involve relatively weaker interactions such as hydrogen bonding and pi-stacking (Matthews, Welton, and Hunt 2015). These additional interactions play a significant role in determining the physical properties of ILs. The forces present in different ILs can vary, leading to variations in their physical properties. The nature of these forces primarily controls properties such as viscosity, melting point, and conductivity (Pei et al. 2022). Understanding the forces at play in ILs is crucial for tailoring their properties for specific applications. The structure of ILs is a key factor in determining their properties. The arrangement of ions and molecules within an IL affects its overall behavior. For example, the presence of specific functional groups or the size and shape of the ions can influence properties like solubility and stability. Among the different types of ILs, imidazolium-based ILs have emerged as one of the most widely studied and utilized systems. The structural features of imidazolium-based ILs can be tuned by the substitution of different alkyl groups at the nitrogen atom of the imidazole ring. This substitution can significantly affect the physicochemical properties of the IL, such as its melting point, viscosity, and solubility, as well as its interactions with other molecules (Anthony et al. 2003). The introduction of longer alkyl chains in the imidazolium cation can increase the hydrophobicity of the IL, leading to stronger interactions with nonpolar molecules. On the other hand, shorter alkyl chains can enhance the IL's solubility in polar solvents and weaken its interactions with nonpolar species. In addition, the substitution of different alkyl groups can also affect the packing and ordering of the IL molecules, leading to changes in their liquid crystal behavior and self-assembly properties. This can have important implications for the design and development of IL-based materials for various applications. Overall, the variation of structural features and interactions with the substitution of different alkyls for imidazolium-based ILs is an important area of research that has the potential to lead to the discovery of novel IL-based materials with tailored properties and functions.

1.1. Interaction in Ionic Liquids

Interaction in ionic liquids refers to the various ways in which these unique solvents interact with other substances. Ionic liquids are composed of ions, which are charged particles, and as a result, they have distinct properties that set them apart from traditional solvents. One key aspect of the interaction in ionic liquids is their ability to dissolve a wide range of substances. Due to their charged nature, ionic liquids have a strong affinity for polar and charged molecules. This allows them to dissolve salts, acids, bases, and even some metal complexes. The dissolution process involves the ions of the liquid surrounding and solvating the solute particles, effectively separating them and allowing for homogeneous mixing (Rezabal and Schäfer 2015). Another important aspect of the interaction in ionic liquids is their unique ability to stabilize reactive species. The strong electrostatic forces between the ions in the liquid can stabilize charged intermediates, such as ions or radicals, which are typically highly reactive and short-lived in other solvents. This property has led to the use of ionic liquids as reaction media in various organic and inorganic reactions, including catalysis and electrochemistry. In addition to their ability to dissolve and stabilize substances, ionic liquids also exhibit specific interactions with surfaces. When in contact with solid materials, the ions of the liquid can adsorb onto the surface, forming an ordered layer known as an "ionic liquid film." This film can modify the surface properties, such as wettability, and affect the behavior of the material. These interactions have implications in fields such as tribology, where the lubricating properties of ionic liquids can be exploited (Horvath, Anaredy, and Shaw 2022). Furthermore, the interaction in ionic liquids is influenced by their unique physicochemical properties. For instance, the high viscosity of some ionic liquids can hinder molecular diffusion and affect the rates of chemical reactions. The polarity and charge distribution of the ions also plays a role in their interactions with other substances. Understanding the interaction in ionic liquids is crucial for the design and optimization of processes involving these solvents. By studying and manipulating the interactions, researchers can develop new applications for ionic liquids in various fields, such as energy storage, separation processes, and green chemistry (J. N. Israelachvili 2011). In conclusion, the interaction in ionic liquids encompasses their ability to dissolve substances, stabilize reactive species, interact with surfaces, and influence their unique physicochemical properties. The study of these interactions provides insights into the behavior of ionic liquids and enables the development of innovative applications (Welton 1999) (Rogers and Seddon 2003).

1.2. DFT on Ionic Liquids

DFT (Density Functional Theory) is a computational method used to study the electronic structure and properties of materials. In recent years, it has been widely applied to investigate the behavior of ionic liquids. DFT calculations have proven to be a valuable tool in understanding the fundamental properties of ionic liquids. By employing DFT, researchers can determine the optimized geometries, vibrational frequencies, electronic structures, and energetics of these systems. This information provides insights into the intermolecular interactions, solvation behavior, and transport properties of ionic liquids (Zorn, Boatz, and Gordon 2006).

DFT provides a powerful tool to investigate the behavior of molecules and solid-state systems by solving the Schrödinger equation for the electronic wavefunction. Various theoretical methods such as Semi-empirical, HF, DFT, MP2, MP3, and MP4 are available to carry out different types of calculations in the ground state. All these calculations are based on the solution of the Schrodinger wave equation (1):

$$\hat{H}\Psi\,(r_1, r_2 \ldots\ldots r_N) = E\Psi(r_1, r_2 \ldots\ldots r_N) \qquad (1)$$

where H is the Hamiltonian and is represented as –

$$\hat{H} = \frac{\hbar^2}{2m}\sum_i \nabla_i^2 - \sum_{i,I}\frac{Z_1 e^2}{|r_i - R_I|} + \frac{1}{2}\sum_{i,j}\frac{e^2}{|r_i - r_j|} - \frac{\hbar^2}{2M}\sum_i \nabla_i^2 + \frac{1}{2}\sum_{i,j}\frac{Z_I Z_J e^2}{|R_I - R_J|} \qquad (2)$$

$\Psi(r_1,r_2,\ldots\ldots r_N)$ is the electron wave function and is a function of 3N variables, where N is the number of electrons in the system. The wave function is written to be an antisymmtrized product (Slater Determinant) as shown in equation 3. Ψ as such has no meaning but Ψ^2 is the probability of finding the electron at a particular place. E is the energy of the system.

$$\Psi = \frac{1}{\sqrt{N!}} \begin{vmatrix} \Psi_1(r_1) & \Psi_1(r_2)\cdots & \Psi_1(r_N) \\ \Psi_2(r_1) & \Psi_2(r_2)\cdots & \Psi_2(r_N) \\ \vdots & \vdots \ddots & \vdots \\ \Psi_N(r_1) & \Psi_N(r_2) & \Psi_N(r_N) \end{vmatrix} \quad (3)$$

Semi-empirical and HF methods are lower-level calculations and often provide absurd results though they are quite time-economic. This is sometimes because the HF method ignores the electron correlation in the molecules while performing calculations, thereby showing large deviations from the experimental results as shown in much of the literature (M. Shukla et al. 2014). Hohenberg-Kohn-Sham proposed a new approach to the many-body interacting electron problems where all ground state properties are determined by the ground state density and this calculation is well-known as Density Functional Theory (DFT) calculation. The achievement of DFT not only encompasses standard bulk materials but also complex materials such as proteins (Vila and Scheraga 2009) or carbon nanotubes (Mahdy 2015) With respect to room temperature ionic liquids (RTILs), DFT calculation was found to be very useful in predicting the structure (Ozawa et al. 2003) (Hamaguchi and Ozawa 2005) (Fumino, Wulf, and Ludwig 2009). DFT calculation also helps to understand the interactions present among cations and anions in the molecules as well as the type of bonding present in the molecule (Fumino, Wulf, and Ludwig 2009) (Wulf, Fumino, and Ludwig 2010) (Kempter and Kirchner 2010). Several other quantum chemical calculations have been found to predict the melting point, dielectric constant, Gibbs free energy, enthalpy, and several other thermodynamic properties reasonably well (Fumino et al. 2011) (Li et al. 2011).

2. Key Aspects of DFT Calculations on Ionic Liquids

2.1. Structure Optimization

Quantum chemical methods have gained significant attention in the study of ionic liquids due to their ability to provide precise energetic information about the interplay of fundamental forces. These forces include electrostatics, induction, exchange, and dispersion. By employing quantum chemical methods, we can investigate the intricate details of the molecular interactions within ionic liquids. The accurate energetic insights obtained through these methods allow us to understand how electrostatics and dispersion forces influence the behavior and properties of these liquids. One of the primary objectives of DFT calculations on ionic liquids is to determine their optimized molecular or crystal structures (Liu, pu, and Chen 2005; Adhikari et al. 2023). By minimizing the total energy of the system, the most stable arrangement of ions can be obtained. This allows researchers to study the packing arrangements, intermolecular interactions, and the influence of different cations and anions on the overall structure. Several literatures are available explaining the molecular structure determination of ILs using DFT calculations (Seeger and Izgorodina 2020), (M. Shukla 2017; M. Shukla, Srivastava, and Saha, n.d.). Also, the different types of interactions present in ILs can be identified using DFT calculations (M. Shukla, Srivastava, and Saha 2010) (M. Shukla et al. 2014). Seeger et al. has recently performed DFT and MP2 calculations on cluster of two ion pairs of imidazolium ILs with 43 different levels of theory. Combinations of density functional theory (DFT) functionals and basis sets have been extensively studied by them to optimize the geometry of ion-paired clusters in imidazolium ionic liquids. Several combinations, including ωB97X-D/cc-pVDZ, M06-2X/aug-cc-pVDZ, B3LYP-D3/cc-pVTZ, and TPSS-D3/cc-pVTZ, have shown excellent performance in optimizing the geometry of two ion-paired clusters of imidazolium ionic liquids. However, when these same combinations were applied to four ion-paired clusters composed of varying ionic liquids, larger deviations were observed. The optimized geometries deviated from the experimental data, indicating that these combinations may not be as suitable for this particular set of ion-paired clusters (Seeger and Izgorodina 2020). Hence using the suitable method and basis set, very accurate structure can be analyzed. Their energies and interactions can be analyzed well which is difficult to analyze experimentally.

2.2. Vibrational Spectroscopy

DFT calculations are a powerful tool in computational chemistry that can be utilized to determine the vibrational properties of ionic liquids. These properties include the vibrational frequencies, intensities, and modes of vibration. Vibrational frequencies represent the energies associated with the different modes of vibration in the system, while intensities indicate the relative strengths of these vibrations. The modes of vibration describe the specific motion of atoms within the ionic liquid. By determining the vibrational properties of ionic liquids using DFT calculations, researchers can gain insights into the behavior and properties of these materials. DFT calculations can also be used to determine the vibrational properties of ionic liquids, providing insights into their dynamic behavior and intermolecular interactions (Drai et al. 2017). By calculating the vibrational frequencies and corresponding infrared (IR) or Raman spectra, researchers can identify characteristic vibrational modes and study how they are affected by changes in the ionic liquid's composition or temperature (M. Shukla et al. 2014; M. Shukla, Srivastava, and Saha 2010; n.d.). Several literatures are available explaining that DFT help to analyse the new peaks in novel compounds as well as correlates experimental results to a greater extent (Guglielmero et al. 2019) (Paschoal, Faria, and Ribeiro 2017) (Mondal and Balasubramanian 2015).

2.3. Thermodynamic Properties

Thermodynamic properties play a crucial role in understanding and predicting the behavior of chemical systems. The Gibbs free energy, enthalpy, and entropy are fundamental thermodynamic properties that provide valuable insights into the stability, energetics, and spontaneity of chemical reactions. In recent years, the application of Density Functional Theory (DFT) has emerged as a powerful tool for calculating these properties, especially for ionic liquids. The accurate prediction of thermodynamic properties for ionic liquids is crucial for the design and optimization of processes involving these versatile materials. Traditionally, experimental methods, such as calorimetry and spectroscopy, have been employed to measure these properties. However, these experimental techniques can be time-consuming, expensive, and sometimes challenging to perform for ionic liquids due to their unique nature. Here DFT comes into play. By solving the electronic Schrödinger equation, DFT can provide valuable insights into various properties, including thermodynamic properties, of ionic liquids. To calculate thermodynamic properties using DFT, one starts by constructing an appropriate model of the ionic liquid system, considering the interactions between ions and the surrounding solvent molecules. These models can range from simple cluster models to more complex periodic boundary conditions, depending on the level of accuracy desired. Once the model is established, DFT calculations can be performed to obtain the electronic structure and energies of the system. From these electronic energies, thermodynamic properties such as the Gibbs free energy, enthalpy, and entropy can be determined using statistical thermodynamics.

It is important to note that accurate calculations of thermodynamic properties require the consideration of various factors, such as temperature, pressure, and solvation effects. Additionally, the choice of exchange-correlation functional and basis set in DFT calculations can significantly impact the accuracy of the results. Hence thermodynamic properties, such as the Gibbs free energy, entropy and enthalpy can be calculated using DFT. These calculations enable researchers to predict phase behavior, solvation energies, and reaction energetics of ionic liquids. By comparing the thermodynamic properties of different ionic liquids, one can gain a better understanding of their stability and reactivity (Karu et al. 2016) (M. Shukla and Saha 2013). In conclusion, Density Functional Theory (DFT) has proven to be a valuable tool for calculating thermodynamic properties, including the Gibbs free energy, enthalpy, and entropy, for ionic liquids (Glasser and Jenkins 2016) (Karu et al. 2016). By providing insights into the stability and energetics of these materials, DFT calculations can aid in the design and optimization of processes involving ionic liquids, contributing to advancements in various fields of science and technology.

2.4. Electrochemical Properties

One of the key advantages of imidazolium ionic liquids is their excellent electrochemical stability. Unlike traditional organic solvents, which can decompose or react with electrode surfaces, imidazolium ionic liquids remain stable over a wide range of potentials. This stability makes them ideal for use as electrolytes in electrochemical cells and devices. Another important characteristic of imidazolium ionic liquids is their high ionic conductivity. Due to the presence of charged species, these liquids exhibit excellent transport properties, enabling efficient ion transfer between electrodes. This high conductivity is crucial for the performance of electrochemical devices, such as batteries and supercapacitors. Furthermore, imidazolium ionic liquids possess a wide electrochemical window, meaning they can withstand high voltages without undergoing significant degradation. This property allows for the exploration of new electrochemical processes and the development of novel energy storage systems. In addition to their stability and conductivity, imidazolium ionic liquids also offer tunable properties. By varying the structures of both the cations and anions, researchers can customize the physicochemical characteristics of these liquids, such as viscosity, melting point, and solubility (Srivastava, Shukla, and Saha 2010) (H. Wang et al. 2015). This tunability makes them versatile for a wide range of electrochemical applications. Imidazolium ionic liquids have been successfully applied in various electrochemical systems. For example, they have been used as electrolytes in lithium-ion batteries, where they can improve the safety, stability, and performance of the batteries (Seungmin Oh, Keating, and Biddinger 2022). Additionally, these liquids have shown promise in fuel cells, electroplating, and electrochemical sensors. Imidazolium ionic liquids have shown great promise in electrochemical applications. Their excellent ionic conductivity, wide electrochemical window, and stability make them ideal electrolytes for batteries, fuel cells, and supercapacitors (Sato, Masuda, and Takagi 2004). Imidazolium-based electrolytes have been utilized to enhance the performance and stability of lithium-ion batteries, resulting in increased energy density and longer cycle life. Furthermore, these ionic liquids have been explored as electrolytes in dye-sensitized solar cells and electrochemical sensors, showcasing their potential in renewable energy and analytical chemistry (Shah, An, and Muhammad 2020). Hence ILs have been widely used as electrolytes in various electrochemical devices. DFT calculations can provide valuable insights into their electrochemical behavior, including the redox potentials, charge transfer mechanisms, and reaction kinetics. These calculations help in designing novel ionic liquids with improved electrochemical performance. In conclusion, the use of imidazolium ionic liquids in electrochemical applications has demonstrated great promise. Their stability, high conductivity, wide electrochemical window, and tunable properties make them attractive for a range of applications in energy storage, sensing, and electrocatalysis (Cruz and Ciach 2021). Continued research and development in this field will undoubtedly uncover new opportunities for the utilization of imidazolium ionic liquids in various electrochemical devices and technologies.

3. Application of Ionic Liquids

3.1. Solvent Properties

Imidazolium ILs have gained significant attention in recent years due to their unique solvent properties which makes them suitable for various applications.

Their low volatility, high polarity, and wide liquid temperature range make them ideal solvents for organic synthesis, catalysis, and extraction processes (J. K. Singh et al. 2018). They can serve as reaction media, facilitating the synthesis of various compounds, including pharmaceuticals, fine chemicals, and polymers. The unique solvation properties of imidazolium ionic liquids allow for enhanced reaction rates, selectivity, and yields (Veldhorst, Faria, and Ribeiro 2016) (Chang et al. 2018). Additionally, these ionic liquids can be easily modified by changing the cation or anion, enabling tailored solvent properties for specific reactions. These ILs are composed of imidazolium cations, which are organic salts, and various anions. Solvation property of ILs is mainly attributed to the presence of charged ions, which facilitate the dissolution process through ionic interactions. The solvation power of imidazolium ILs can be further enhanced by modifying the cationic or anionic components, enabling tailoring of their solubility for specific applications (Sooyeoun Oh et al. 2012).

However, understanding their solvation behavior is crucial for optimizing their performance in various processes. By employing DFT calculations, researchers can simulate the solvation process of an imidazolium ionic liquid by considering the interactions between the solute molecules and the surrounding solvent molecules (Fatima et al. 2020). These calculations take into account the electronic structure of the molecules and provide valuable information about the energy profiles, molecular geometries, and thermodynamic properties (Palumbo et al. 2021). DFT studies can predict the solvation behavior of imidazolium ionic liquids and provide information about their solute-solvent interactions which are essential for understanding the stability and reactivity of imidazolium ionic liquids in various applications. These interactions can be classified into various types, including hydrogen bonding, electrostatic interactions, and van der Waals forces (Palumbo et al. 2021). Through DFT calculations, researchers can determine the strength and nature of these interactions, which play a crucial role in the solvation behavior of these liquids. Furthermore, DFT studies allow for the investigation of the effects of different solvents on the solvation behavior of imidazolium ionic liquids (Mondal and Balasubramanian 2014) (Jesus et al. 2017). By comparing the solvation energies and structural properties of different solvents, researchers can gain insights into solvent preference and the underlying factors that influence solute-solvent interactions. In conclusion, DFT studies are a powerful tool for predicting the solvation behavior of imidazolium ionic liquids. These studies provide valuable information about the solute-solvent interactions, enabling researchers to understand the underlying mechanisms and optimize the performance of these liquids in various applications.

3.2. Electrochemical Properties

Imidazolium ionic liquids possess unique electrochemical properties due to the presence of charged species. They exhibit wide electrochemical windows, low conductivity, and high thermal stability, making them promising candidates for energy storage devices, such as batteries and supercapacitors (Mei et al. 2018) (Tiago et al. 2020). Imidazolium ionic liquids have shown great promise in electrochemical applications. Their excellent ionic conductivity, wide electrochemical window, and stability make them ideal electrolytes for batteries, fuel cells, and supercapacitors. Imidazolium-based electrolytes have been utilized to enhance the performance and stability of lithium-ion batteries, resulting in increased energy density and longer cycle life. Furthermore, these ionic liquids have been explored as electrolytes in dye-sensitized solar cells and electrochemical sensors, showcasing their potential in renewable energy and analytical chemistry (V. V. Singh et al. 2012). Imidazolium ionic liquids also serve as effective electrode modifiers, enhancing the performance of electrochemical systems. By forming a stable interfacial layer on the electrode surface, Imidazolium ILs can improve the electrochemical activity, increase the specific surface area, and reduce the charge transfer resistance. This modification can significantly enhance the efficiency and stability of electrochemical reactions, leading to improved device performance. Moreover, the unique physicochemical properties of imidazolium cations allow for tailored electrode modifications, enabling the design of electrodes with specific functionalities and selectivity (Sowmiah et al. 2009). DFT calculations can elucidate the redox behavior and stability of imidazolium ionic liquids in various electrochemical applications (Y.-L. Wang et al. 2020).

3.3. Catalysis

Ionic liquids have shown great potential as catalysts in organic synthesis due to their ability to dissolve a wide range of reactants and stabilize reactive intermediates. The use of ionic liquids as catalysts promotes greener and more sustainable processes (Zhao et al. 2002; Maciejewski 2021; Steinrück and

Wasserscheid 2015). Their low volatility reduces the release of harmful emissions into the environment, while their recyclability allows for the recovery and reuse of catalysts, minimizing waste generation. Ionic liquids can serve as catalysts in both homogeneous and heterogeneous reactions (Bartlewicz et al. 2020). In homogeneous catalysis, ionic liquids act as solvents and catalysts simultaneously, providing a unique environment for catalytic transformations. In heterogeneous catalysis, ionic liquids can be immobilized on solid supports, creating supported ionic liquid phase (SILP) catalysts (Lemus et al. 2011) (Latos, Wolny, and Chrobok 2023). SILP catalysts combine the advantages of both homogeneous and heterogeneous catalysis, enabling efficient reactions with easy separation and recycling of the catalyst. Imidazolium ionic liquids have found applications as reaction media and catalysts in various catalytic processes. They can act as solvents for homogeneous catalysts, enabling efficient and selective reactions. Additionally, imidazolium-based catalysts have been developed for a wide range of transformations, such as cross-coupling reactions, oxidation reactions, and polymerization reactions (McLachlan et al. 2003). These catalysts offer advantages such as high stability, recyclability, and tunable catalytic activity, making them attractive for industrial applications (Prechtl, Scholten, and Dupont 2010). The catalytic application of ionic liquids holds great promise for enhancing the efficiency, selectivity, and sustainability of various chemical reactions. Their unique properties, tunability, and ability to function as solvents and catalysts make them valuable tools in organic synthesis, green chemistry, energy conversion, and more.

3.4. Separation Technology

Ionic liquids have gained significant attention in the field of separation technology due to their unique properties and versatile applications which allow for their application in both high-temperature and low-temperature processes (Berthod, Ruiz-Angel, and Carda-Broch 2008) (Ventura et al. 2017). ILs exhibit low vapor pressure, making them less volatile compared to conventional solvents. This characteristic reduces the risk of emissions and makes them environmentally friendly alternatives. The low volatility of ionic liquids leads to reduced energy consumption during separation processes, making them more energy-efficient options. Imidazolium ionic liquids have shown promise in separation technologies due to their unique solvation properties and selectivity. They have been employed in liquid-liquid extraction, gas separation, and membrane technologies. Imidazolium-based solvents can selectively extract specific compounds from complex mixtures, offering an alternative to traditional organic solvents. Moreover, these ionic liquids can be immobilized on solid supports to create functionalized membranes with enhanced separation capabilities (Bento et al. 2021) (Friess et al. 2021). In addition to their solvating properties, ionic liquids can also be modified to have specific functionalities, enhancing their performance in separation processes. By introducing specific functional groups to the ionic liquid structure, researchers can tailor their properties to suit specific separation requirements (Vekariya 2017). This tunability allows for the design of customized ionic liquids for various separation applications.

3.5. Gas Separation and Absorption

Imidazolium ionic liquids have emerged as promising materials in the field of gas separation and absorption (Yang et al. 2022) (Lian et al. 2021). These liquids, composed of imidazolium cations and various anions, have demonstrated exceptional capabilities in effectively separating and absorbing gases. One of the key advantages of imidazolium ionic liquids is their ability to selectively absorb specific gases, allowing for efficient separation of gas mixtures. Through careful selection of the anion component, the properties of the ionic liquid can be tailored to target specific gases. This makes imidazolium ionic liquids highly versatile and adaptable for various gas separation processes (Lian et al. 2021).

Imidazolium ionic liquids have shown remarkable absorption capabilities for a wide range of gases, including carbon dioxide (CO2) and sulfur dioxide (SO2) (L. Wang et al. 2020) (Ramdin, de Loos, and Vlugt 2012). The strong interaction between the imidazolium cation and the gas molecules leads to high absorption capacities and selectivities. This makes them particularly suitable for applications such as carbon capture and storage, where the removal of CO2 from gas streams is crucial. Furthermore, imidazolium ionic liquids have been found to exhibit excellent stability and low volatility, making them suitable for long-term use in gas separation processes. Their thermal stability and non-flammability properties ensure safe operation, while their low vapor pressure minimizes the risk of emissions and loss of valuable gases. In addition to their gas absorption capabilities, imidazolium ionic liquids have also shown the potential in enhancing gas permeation through membranes. By incorporating these liquids into membrane materials, the gas permeability and selectivity can be significantly improved. This opens up possibilities for the development of more efficient gas separation membranes for applications such as gas purification and natural gas processing.

In summary, imidazolium ionic liquids have demonstrated remarkable gas separation and absorption capabilities. Their selective absorption properties, stability, and low volatility make them attractive materials for various applications in the field of gas separation and absorption. Imidazolium ionic liquids have shown remarkable gas separation and absorption capabilities. Their selective gas absorption and high gas solubility make them suitable for carbon capture, natural gas purification, and gas separation processes. DFT studies can provide insights into the interactions between imidazolium ionic liquids and different gas molecules, aiding in the design of efficient gas separation systems (Yu et al. 2021).

3.6. Biological Applications

Imidazolium ionic liquids have also found applications in various biological fields, including biomolecule stabilization, enzyme catalysis, and drug delivery (Egorova, Gordeev, and Ananikov 2017) (Correia et al. 2021). Their ability to interact with biomolecules and their low toxicity make them attractive for biomedical applications. One of the key areas where imidazolium ionic liquids have made an impact is in biocatalysis. Due to their unique properties, such as high thermal stability and low vapor pressure, these liquids have been utilized as solvents for various biocatalytic reactions. They offer advantages over traditional organic solvents by providing a more stable environment for enzymes, improving their catalytic efficiency, and enabling the use of enzymes in non-aqueous systems (Pedro et al. 2020). Imidazolium ionic liquids have also been explored for their antimicrobial properties (Nikfarjam et al. 2021). Studies have shown that certain imidazolium-based compounds possess potent antimicrobial activity against a wide range of microorganisms, including bacteria, fungi, and viruses (Curreri, Mitragotri, and Tanner 2021). This antimicrobial activity makes them promising candidates for the development of new antimicrobial agents, with potential applications in healthcare, food safety, and hygiene. In addition, imidazolium ionic liquids have been investigated for their ability to enhance drug delivery systems. These liquids can act as solvents or carriers for drugs, facilitating their transport across biological barriers and improving their bioavailability. Their unique physicochemical properties, such as their ability to dissolve a wide range of compounds and their tunable hydrophobicity, enable them to overcome some of the limitations associated with traditional drug delivery systems. Furthermore, imidazolium ionic liquids have been utilized in the field of bioseparation. These liquids have shown promise as alternative solvents for the extraction and purification of biomolecules, such as proteins and nucleic acids (S. K. Shukla and Mikkola 2020). Their ability to selectively extract target molecules from complex mixtures, along with their tunable properties, makes them valuable tools in various bioseparation processes. In summary, imidazolium ionic liquids have emerged as versatile compounds with significant applications in various biological fields. Their unique properties and potential for customization make them attractive candidates for a wide range of biomedical and biotechnological applications. Continued research and development in this area are expected to uncover further opportunities for the use of imidazolium ionic liquids in advancing biological sciences.

DFT calculations can help understand the interactions between imidazolium ionic liquids and biomolecules, facilitating the design of novel drug delivery systems and therapeutic agents (Abrar Siddiquee et al. 2023) (Kumar et al. 2020).

Conclusion

Density Functional Theory calculations have proven to be a valuable tool in studying the properties and behavior of ionic liquids. By providing a detailed understanding of their structure, vibrational properties, thermodynamics, and electrochemical behavior, DFT calculations contribute to the development and optimization of ionic liquids for various applications. Continued advancements in computational methods and increased computational power will further enhance the accuracy and applicability of DFT calculations in this field. Further research and exploration in this field hold immense potential for the discovery of new and innovative applications of imidazolium ionic liquids.

Disclaimer

None.

References

Abrar Siddiquee, Md., Juhi Saraswat, Mehraj ud din Parray, Prashant Singh, Savita Bargujar, and Rajan Patel. 2023. "Spectroscopic and DFT Study of Imidazolium Based Ionic Liquids with Broad Spectrum Antibacterial Drug Levofloxacin." *Spectrochimica Acta Part A: Molecular and Biomolecular Spectroscopy* 285: 121803. https://doi.org/https://doi.org/10.1016/j.saa.2022.121803.

Adhikari, Rajan, Stephen Massicot, Lukas Fromm, Timo Talwar, Afra Gezmis, and Manuel Meusel. 2023. "Structure and Reactivity of the Ionic Liquid -[C 1 C 1 Im][Tf 2 N] on Cu(111)," no. 111: 22–24. https://doi.org/10.1007/s11244-023-01801-y.

Anthony, Jennifer, Joan Brennecke, Jd Holbrey, Edward Maginn, Rob Mantz, Robin Rogers, Paul Trulove, Ann Visser, and Tom Welton. 2003. "Physicochemical Properties of Ionic Liquids." In, 41–126. https://doi.org/10.1002/3527600701.ch3.

Bartlewicz, Olga, Izabela Dąbek, Anna Szymańska, and Hieronim Maciejewski. 2020. "Heterogeneous Catalysis with the Participation of Ionic Liquids." *Catalysts*. https://doi.org/10.3390/catal10111227.

Bento, Rui MF, Catarina AS Almeida, Márcia C Neves, Ana PM Tavares, and Mara G Freire. 2021. "Advances Achieved by Ionic-Liquid-Based Materials as Alternative Supports and Purification Platforms for Proteins and Enzymes." *Nanomaterials (Basel, Switzerland)* 11 (10). https://doi.org/10.3390/nano11102542.

Berthod, A, MJ Ruiz-Angel, and S Carda-Broch. 2008. "Ionic Liquids in Separation Techniques." *Journal of Chromatography. A* 1184 (1–2): 6–18. https://doi.org/10.1016/j.chroma.2007.11.109.

Chang, Jui-Cheng, Che-Hsuan Yang, I-Wen Sun, Wen-Yueh Ho, and Tzi-Yi Wu. 2018. "Synthesis and Properties of Magnetic Aryl-Imidazolium Ionic Liquids with Dual Brønsted/Lewis Acidity." *Materials (Basel, Switzerland)* 11 (12). https://doi.org/10.3390/ma11122539.

Correia, Daniela Maria, Liliana Correia Fernandes, Margarida Macedo Fernandes, Bruno Hermenegildo, Rafaela Marques Meira, Clarisse Ribeiro, Sylvie Ribeiro, Javier Reguera, and Senentxu Lanceros-Méndez. 2021. "Ionic Liquid-Based Materials for Biomedical Applications." *Nanomaterials (Basel, Switzerland)* 11 (9). https://doi.org/10.3390/nano11092401.

Cruz, Carolina, and Alina Ciach. 2021. "Phase Transitions and Electrochemical Properties of Ionic Liquids and Ionic Liquid-Solvent Mixtures." *Molecules (Basel, Switzerland)* 26 (12). https://doi.org/10.3390/molecules26123668.

Curreri, Alexander M, Samir Mitragotri, and Eden EL Tanner. 2021. "Recent Advances in Ionic Liquids in Biomedicine." *Advanced Science* 8 (17): 1–18. https://doi.org/10.1002/advs.202004819.

Drai, M, A Mostefai, A Paolone, B Haddad, E Belarbi, D Villemin, S Bresson, O Abbas, Y Chaker, and M Rahmouni. 2017. "Synthesis, Experimental and Theoretical Vibrational Studies Of." *Journal of Chemical Sciences* 129 (6): 707–19. https://doi.org/10.1007/s12039-017-1282-6.

Egorova, Ksenia S, Evgeniy G Gordeev, and Valentine P Ananikov. 2017. "Biological Activity of Ionic Liquids and Their Application in Pharmaceutics and Medicine." *Chemical Reviews* 117 (10): 7132–89. https://doi.org/10.1021/acs.chemrev.6b00562.

Fatima, Urooj, Riyazuddeen, Mohammad Jane Alam, and Shabbir Ahmad. 2020. "Experimental Thermophysical Properties and DFT Calculations of Imidazolium Ionic Liquids and 2-Butanol Mixtures." *Fluid Phase Equilibria* 508: 112447. https://doi.org/https://doi.org/10.1016/j.fluid.2019.112447.

Friess, Karel, Pavel Izák, Magda Kárászová, Mariia Pasichnyk, Marek Lanč, Daria Nikolaeva, Patricia Luis, and Johannes C Jansen. 2021. "A Review on Ionic Liquid Gas Separation Membranes." *Membranes*. https://doi.org/10.3390/membranes 11020097.

Fumino, Koichi, Tim Peppel, Monika Geppert-Rybczyńska, Dzmitry H Zaitsau, Jochen K Lehmann, Sergey P Verevkin, Martin Köckerling, and Ralf Ludwig. 2011. "The Influence of Hydrogen Bonding on the Physical Properties of Ionic Liquids." *Physical Chemistry Chemical Physics* 13 (31): 14064–75. https://doi.org/10.1039/C1CP20732F.

Fumino, Koichi, Alexander Wulf, and Ralf Ludwig. 2009. "The Potential Role of Hydrogen Bonding in Aprotic and Protic Ionic Liquids." *Phys. Chem. Chem. Phys.* 11 (39): 8790–94. https://doi.org/10.1039/B905634C.

Glasser, Leslie, and H Donald Brooke Jenkins. 2016. "Liquids †." https://doi.org/10.1039/c6cp00235h.

Guglielmero, Luca, Lorenzo Guazzelli, Alessandra Toncelli, Cinzia Chiappe, Alessandro Tredicucci, and Christian Silvio Pomelli. 2019. "An Insight into the Intermolecular Vibrational Modes of Dicationic Ionic Liquids through Far-Infrared Spectroscopy and DFT Calculations." *RSC Advances* 9 (52): 30269–76. https://doi.org/10.1039/C9RA05735H.

Hamaguchi, Hiro-O., and Ryosuke Ozawa. 2005. "Structure of Ionic Liquids and Ionic Liquid Compounds: Are Ionic Liquids Genuine Liquids in the Conventional Sense?" *ChemInform* 36 (47): 85–104. https://doi.org/10.1002/chin.200547277.

Horvath, Andrew, Radhika S Anaredy, and Scott K Shaw. 2022. "Solvents and Stabilization in Ionic Liquid Films." *Langmuir* 38 (30): 9372–81. https://doi.org/10.1021/acs.langmuir.2c01258.

Israelachvili, Jacob N. 2011. "8 - Special Interactions: Hydrogen-Bonding and Hydrophobic and Hydrophilic Interactions." In: edited by Jacob NBT - Intermolecular and Surface Forces (Third Edition) Israelachvili, 151–67. San Diego: Academic Press. https://doi.org/https://doi.org/10.1016/B978-0-12-375182-9.10008-9.

Jesus, C De, Paulo AR Pires, Rizwana Mustafa, Naheed Riaz, and Omar A El Seoud. 2017. "RSC Advances Experimental and Theoretical Studies on Solvation in Aqueous Solutions of Ionic Liquids Carrying Different Side Chains: The n-Butyl-Group versus the Methoxyethyl Group†," 15952–63. https://doi.org/10.1039/c7ra00273d.

Karu, Karl, Anton Ruzanov, Heigo Ers, Vladislav Ivaništšev, Isabel Lage-Estebanez, and José M García de la Vega. 2016. "Predictions of Physicochemical Properties of Ionic Liquids with DFT." *Computation*. https://doi.org/10.3390/computation4030025.

Kempter, V, and B Kirchner. 2010. "The Role of Hydrogen Atoms in Interactions Involving Imidazolium-Based Ionic Liquids." *Journal of Molecular Structure* 972 (1): 22–34. https://doi.org/https://doi.org/10.1016/j.molstruc.2010.02.003.

Kumar, Pannuru Kiran, Indrani Jha, Anamika Sindhu, Pannuru Venkatesu, Indra Bahadur, and Eno E Ebenso. 2020. "Experimental and Molecular Docking Studies in Understanding the Biomolecular Interactions between Stem Bromelain and Imidazolium-Based Ionic Liquids." *Journal of Molecular Liquids* 297: 111785. https://doi.org/https://doi.org/10.1016/j.molliq.2019.111785.

Latos, Piotr, Anna Wolny, and Anna Chrobok. 2023. "Supported Ionic Liquid Phase Catalysts Dedicated for Continuous Flow Synthesis." *Materials*. https://doi.org/10.3390/ma16052106.

Lemus, Jesus, Jose Palomar, Miguel A Gilarranz, and Juan J Rodriguez. 2011. "Characterization of Supported Ionic Liquid Phase (SILP) Materials Prepared from Different Supports." *Adsorption* 17 (3): 561–71. https://doi.org/10.1007/s10450-011-9327-5.

Li, Wei, Chuan Song Qi, Xin Min Wu, Hua Rong, and Liang Fa Gong. 2011. "Relationship between Melting Point and the Interaction Energy of Alkyl Imidazolium Tetrafluoroborate Ionic Liquids." *Wuli Huaxue Xuebao/ Acta Physico - Chimica Sinica* 27 (9): 2059–64. https://doi.org/10.3866/pku.whxb20110914.

Lian, Shaohan, Chunfeng Song, Qingling Liu, Erhong Duan, Hongwei Ren, and Yutaka Kitamura. 2021. "Recent Advances in Ionic Liquids-Based Hybrid Processes for CO2 Capture and Utilization." *Journal of Environmental Sciences* 99: 281–95. https://doi.org/https://doi.org/10.1016/j.jes.2020.06.034.

Liu, K.-H, Min pu, and Biaohua Chen. 2005. "DFT Study on the Structure of Ionic Liquid 1-Ethyl-3-Methylimidazolium Hexafluorophosphate." *Jiegou Huaxue* 24 (January): 576–80.

Maciejewski, Hieronim. 2021. "Ionic Liquids in Catalysis." *Catalysts*. https://doi.org/10.3390/catal11030367.

Mahdy, A. 2015. "DFT Study of Hydrogen Storage in Pd-Decorated C 60 Fullerene." *Molecular Physics* 113 (April): 1–14. https://doi.org/10.1080/00268976.2015.1039090.

Matthews, Richard, Tom Welton, and Patricia Hunt. 2015. "Hydrogen Bonding and π-π Interactions in Imidazolium-Chloride Ionic Liquid Clusters." *Phys. Chem. Chem. Phys.* 17 (March). https://doi.org/10.1039/C5CP00459D.

McLachlan, Fiona, Christopher J Mathews, Paul J Smith, and Tom Welton. 2003. "Palladium-Catalyzed Suzuki Cross-Coupling Reactions in Ambient Temperature Ionic Liquids: Evidence for the Importance of Palladium Imidazolylidene Complexes." *Organometallics* 22 (25): 5350–57. https://doi.org/10.1021/om034075y.

Mei, Xinyi, Zheng Yue, Qiang Ma, Hamza Dunya, and Braja K Mandal. 2018. "Synthesis and Electrochemical Properties of New Dicationic Ionic Liquids." *Journal of Molecular Liquids* 272: 1001–18. https://doi.org/https://doi.org/10.1016/j.molliq.2018.10.085.

Mondal, Anirban, and Sundaram Balasubramanian. 2014. "Quantitative Prediction of Physical Properties of Imidazolium Based Room Temperature Ionic Liquids through Determination of Condensed Phase Site Charges: A Refined Force Field." *The Journal of Physical Chemistry B* 118 (12): 3409–22. https://doi.org/10.1021/jp500296x.

———. 2015. "Vibrational Signatures of Cation − Anion Hydrogen Bonding in Ionic Liquids: A Periodic Density Functional Theory and Molecular Dynamics Study." https://doi.org/10.1021/jp5113679.

Nikfarjam, N, Ghomi, M, Agarwal, T, Hassanpour, M, Sharifi, E, Khorsandi, D, Ali Khan, M, Rossi, F, Rossetti, A, Nazarzadeh Zare, E, Rabiee, N, Afshar, D, Vosough, M, Kumar Maiti, T, Mattoli, V, Lichtfouse, E, Tay, FR, & Makvandi, P. 2021. "Antimicrobial Ionic Liquid-Based Materials for Biomedical Applications." *Advanced Functional Materials* 31 (42). https://doi.org/10.1002/adfm.202104148.

Oh, Seungmin, Michael J Keating, and Elizabeth J Biddinger. 2022. "Physical and Electrochemical Properties of Ionic-Liquid- and Ester-Based Cosolvent Mixtures with Lithium Salts." *Industrial & Engineering Chemistry Research* 61 (33): 12118–31. https://doi.org/10.1021/acs.iecr.2c00498.

Oh, Sooyeoun, Kang Jeong Won, Byung Park, and Ki-Sub Kim. 2012. "Physical and Thermodynamic Properties of Imidazolium Ionic Liquids." *Korean Chemical Engineering Research* 50 (August): 708–12. https://doi.org/10.9713/kcer.2012.50.4.708.

Ozawa, Ryosuke, Satoshi Hayashi, Satyen Saha, Akiko Kobayashi, and Hiro-o Hamaguchi. 2003. "Rotational Isomerism and Structure of the 1-Butyl-3-Methylimidazolium Cation in the Ionic Liquid State." *Chemistry Letters* 32 (10): 948–49. https://doi.org/10.1246/cl.2003.948.

Palumbo, Oriele, Adriano Cimini, Francesco Trequattrini, Jean-Blaise Brubach, Pascale Roy, and Annalisa Paolone. 2021. "Evidence of the CH···O HydrogenBonding in Imidazolium-Based Ionic Liquids from Far-Infrared Spectroscopy Measurements and DFT Calculations." *International Journal of Molecular Sciences*. https://doi.org/10.3390/ijms22116155.

Paschoal, Vitor H, Luiz FO Faria, and Mauro CC Ribeiro. 2017. "Vibrational Spectroscopy of Ionic Liquids." *Chemical Reviews* 117 (10): 7053–7112. https://doi.org/10.1021/acs.chemrev.6b00461.

Pedro, Sónia N, Carmen SR. Freire, Armando JD Silvestre, and Mara G Freire. 2020. "The Role of Ionic Liquids in the Pharmaceutical Field: An Overview of Relevant Applications." *International Journal of Molecular Sciences*. https://doi.org/10.3390/ijms21218298.

Pei, Yuanchao, Yaxin Zhang, Jie Ma, Maohong Fan, Suojiang Zhang, and Jianji Wang. 2022. "Ionic Liquids for Advanced Materials." *Materials Today Nano* 17: 100159. https://doi.org/https://doi.org/10.1016/j.mtnano.2021.100159.

Prechtl, Martin HG, Jackson D Scholten, and Jairton Dupont. 2010. "Carbon-Carbon Cross Coupling Reactions in Ionic Liquids Catalysed by Palladium Metal Nanoparticles." *Molecules (Basel, Switzerland)* 15 (5): 3441–61. https://doi.org/10.3390/molecules15053441.

Ramdin, Mahinder, Theo W de Loos, and Thijs JH Vlugt. 2012. "State-of-the-Art of CO2 Capture with Ionic Liquids." *Industrial & Engineering Chemistry Research* 51 (24): 8149–77. https://doi.org/10.1021/ie3003705.

Rezabal, Elixabete, and Thomas Schäfer. 2015. "Ionic Liquids as Solvents of Polar and Non-Polar Solutes: Affinity and Coordination." *Phys. Chem. Chem. Phys.* 17 (May). https://doi.org/10.1039/C5CP01774B.

Rogers, Robin D, and Kenneth R Seddon. 2003. "Chemistry. Ionic Liquids--Solvents of the Future?" *Science (New York, N.Y.)* 302 (5646): 792–93. https://doi.org/10.1126/science.1090313.

Sato, Takaya, Gen Masuda, and Kentaro Takagi. 2004. "Electrochemical Properties of Novel Ionic Liquids for Electric Double Layer Capacitor Applications." *Electrochimica Acta* 49 (21): 3603–11. https://doi.org/https://doi.org/10.1016/j.electacta.2004.03.030.

Seeger, Zoe L, and Ekaterina I Izgorodina. 2020. "A Systematic Study of DFT Performance for Geometry Optimizations of Ionic Liquid Clusters." *Journal of Chemical Theory and Computation* 16 (10): 6735–53. https://doi.org/10.1021/acs.jctc.0c00549.

Shah, Faiz Ullah, Rong An, and Nawshad Muhammad. 2020. "Editorial: Properties and Applications of Ionic Liquids in Energy and Environmental Science." *Frontiers in Chemistry* 8. https://doi.org/10.3389/fchem.2020.627213.

Shukla, Madhulata. 2017. "Hydrogen Bonding Interactions in Nicotinamide Ionic Liquids: A Comparative Spectroscopic and DFT Studies." *Journal of Molecular Structure* 1131: 275–80. https://doi.org/https://doi.org/10.1016/j.molstruc.2016.11.067.

Shukla, Madhulata, Hemanth Noothalapati, Shinsuke Shigeto, and Satyen Saha. 2014. "Importance of Weak Interactions and Conformational Equilibrium in N-Butyl-N-Methylpiperidinium Bis(Trifluromethanesulfonyl) Imide Room Temperature Ionic Liquids: Vibrational and Theoretical Studies." *Vibrational Spectroscopy* 75: 107–17. https://doi.org/10.1016/j.vibspec.2014.10.006.

Shukla, Madhulata, and Satyen Saha. 2013. "Relationship between Stabilization Energy and Thermophysical Properties of Different Imidazolium Ionic Liquids: DFT Studies." *Computational and Theoretical Chemistry* 1015: 27–33. https://doi.org/https://doi.org/10.1016/j.comptc.2013.04.007.

Shukla, Madhulata, Nitin Srivastava, and Satyen Saha. n.d. "Interactions and Transitions in Imidazolium Cation Based Ionic Liquids."

———. 2010. "Theoretical and Spectroscopic Studies of 1-Butyl-3-Methylimidazolium Iodide Room Temperature Ionic Liquid: Its Differences with Chloride and Bromide Derivatives." *Journal of Molecular Structure* 975 (1): 349–56. https://doi.org/https://doi.org/10.1016/j.molstruc.2010.05.003.

Shukla, Shashi Kant, and Jyri-Pekka Mikkola. 2020. "Use of Ionic Liquids in Protein and DNA Chemistry." *Frontiers in Chemistry* 8: 598662. https://doi.org/10.3389/fchem.2020.598662.

Singh, Jitendra Kumar, Rahul Kumar Sharma, Pushpal Ghosh, and Ashwani Kumar. 2018. "Imidazolium Based Ionic Liquids : A Promising Green Solvent for Water Hyacinth Biomass Deconstruction" 6 (November). https://doi.org/10.3389/fchem.2018.00548.

Singh, Virendra V, Anil K Nigam, Anirudh Batra, Mannan Boopathi, Beer Singh, and Rajagopalan Vijayaraghavan. 2012. "Applications of Ionic Liquids in Electrochemical Sensors and Biosensors." Edited by Sherif Zein El Abedin. *International Journal of Electrochemistry* 2012: 165683. https://doi.org/10.1155/2012/165683.

Sowmiah, Subbiah, Venkatesan Srinivasadesikan, Ming-Chung Tseng, and Yen-Ho Chu. 2009. "On the Chemical Stabilities of Ionic Liquids." *Molecules*. https://doi.org/10.3390/molecules14093780.

Srivastava, Nitin, Madhulata Shukla, and Satyen Saha. 2010. "An Unusal Effect of Charcoal on the Purification of Alkylimidazolium Iodide Room Temperature Ionic Liquids." *Indian Journal of Chemistry - Section A Inorganic, Physical, Theoretical and Analytical Chemistry* 49 (5–6): 757–61.

Steinrück, Hans-Peter, and Peter Wasserscheid. 2015. "Ionic Liquids in Catalysis." *Catalysis Letters* 145 (1): 380–97. https://doi.org/10.1007/s10562-014-1435-x.

Tiago, Gonçalo AO, Inês AS Matias, Ana PC Ribeiro, and Luísa MDRS Martins. 2020. "Application of Ionic Liquids in Electrochemistry-Recent Advances." *Molecules (Basel, Switzerland)* 25 (24). https://doi.org/10.3390/molecules25245812.

Vekariya, Rohit L. 2017. "A Review of Ionic Liquids: Applications towards Catalytic Organic Transformations." *Journal of Molecular Liquids* 227: 44–60. https://doi.org/https://doi.org/10.1016/j.molliq.2016.11.123.

Veldhorst, Arno A, Luiz FO Faria, and Mauro CC Ribeiro. 2016. "Local Solvent Properties of Imidazolium-Based Ionic Liquids." *Journal of Molecular Liquids* 223: 283–88. https://doi.org/https://doi.org/10.1016/j.molliq.2016.08.044.

Ventura, Sónia PM, Francisca A e Silva, Maria V Quental, Dibyendu Mondal, Mara G Freire, and João AP Coutinho. 2017. "Ionic-Liquid-Mediated Extraction and Separation Processes for Bioactive Compounds: Past, Present, and Future Trends." *Chemical Reviews* 117 (10): 6984–7052. https://doi.org/10.1021/acs.chemrev.6b00550.

Vila, Jorge A, and Harold A Scheraga. 2009. "Assessing the Accuracy of Protein Structures by Quantum Mechanical Computations of 13C(Alpha) Chemical Shifts." *Accounts of Chemical Research* 42 (10): 1545–53. https://doi.org/10.1021/ar900068s.

Wang, Huali, Sichen Gu, Ying Bai, Shi Chen, Na Zhu, Chuan Wu, and Feng Wu. 2015. "Anion-Effects on Electrochemical Properties of Ionic Liquid Electrolytes for Rechargeable Aluminum Batteries." *Journal of Materials Chemistry A* 3 (45): 22677–86. https://doi.org/10.1039/C5TA06187C.

Wang, Lanyun, Yajuan Zhang, Yang Liu, Huilong Xie, Yongliang Xu, and Jianping Wei. 2020. "SO2 Absorption in Pure Ionic Liquids: Solubility and Functionalization." *Journal of Hazardous Materials* 392: 122504. https://doi.org/https://doi.org/10.1016/j.jhazmat.2020.122504.

Wang, Yong-Lei, Bin Li, Sten Sarman, Francesca Mocci, Zhong-Yuan Lu, Jiayin Yuan, Aatto Laaksonen, and Michael D Fayer. 2020. "Microstructural and Dynamical Heterogeneities in Ionic Liquids." *Chemical Reviews* 120 (13): 5798–5877. https://doi.org/10.1021/acs.chemrev.9b00693.

Wang, Zhe, Shuangding He, Vincent Nguyen, and Kevin E Riley. 2020. "Ionic Liquids as “Green Solvent and/or Electrolyte” for Energy Interface ." *Engineered Science* . https://doi.org/10.30919/es8d0013.

Welton, Thomas. 1999. "Room-Temperature Ionic Liquids. Solvents for Synthesis and Catalysis." *Chemical Reviews* 99 (8): 2071–84. https://doi.org/10.1021/cr980032t.

———. 2002. *Chemical Synthesis Using Supercritical Fluids Organic Synthesis on Solid Phase Microwaves in Organic Synthesis Solvent-Free Organic Synthesis*. Vol. 7.

Wulf, Alexander, Koichi Fumino, and Ralf Ludwig. 2010. "Spectroscopic Evidence for an Enhanced Anion-Cation Interaction from Hydrogen Bonding in Pure Imidazolium Ionic Liquids." *Angewandte Chemie - International Edition* 49 (2): 449–53. https://doi.org/10.1002/anie.200905437.

Yang, Bingbing, Haiyan Jiang, Lu Bai, Yinge Bai, Ting Song, and Xiangping Zhang. 2022. "Application of Ionic Liquids in the Mixed Matrix Membranes for CO2 Separation: An Overview." *International Journal of Greenhouse Gas Control* 121: 103796. https://doi.org/https://doi.org/10.1016/j.ijggc.2022.103796.

Yu, Gangqiang, Ruinian Xu, Bin Wu, Ning Liu, Biaohua Chen, Chengna Dai, Yu Kuang, and Zhigang Lei. 2021. "Molecular Thermodynamic and Dynamic Insights into Gas Dehydration with Imidazolium–Based Ionic Liquids." *Chemical Engineering Journal* 416: 129168. https://doi.org/https://doi.org/10.1016/j.cej.2021.129168.

Zhao, Dongbin, Min Wu, Yuan Kou, and Enze Min. 2002. "Ionic Liquids: Applications in Catalysis." *Catalysis Today* 74 (1): 157–89. https://doi.org/https://doi.org/10.1016/S0920-5861(01)00541-7.

Zorn, Deborah, Jerry Boatz, and Mark Gordon. 2006. "Electronic Structure Studies of Tetrazolium-Based Ionic Liquids." *The Journal of Physical Chemistry. B* 110 (July): 11110–19. https://doi.org/10.1021/jp060854r.

Biographical Sketch

Madhulata Shukla

Affiliation: G. B. College Ramgarh, Veer Kunwar Singh University, Bihar, India

Education: PhD

Business Address: Department of Chemistry, G. B. College Ramgarh, Veer Kunwar Singh University, Kaimur, Bihar, India

Research and Professional Experience: Completed Ph.D from Banaras Hindu University in 2014 Post-doctorate from Indian Institute of Technology (Banaras Hindu University) in 2017 and working as assistant professor in Department of Chemistry, G. B. College Ramgarh, Veer Kunwar Singh University, Bihar from 2017.

Professional Appointments: Assistant Professor

Publications from the Last 3 Years: 8 research papers in reputed international journals, 4 book chapters and 1 book edited.

Chapter 5

Bisimidazolium Salts As Ionic Liquid Crystals

Monica Iliş
and Viorel Cîrcu*
Department of Inorganic and Organic Chemistry, Biochemistry and Catalysis, Laboratory of Liquid Crystals, University of Bucharest, Bucharest, Romania

Abstract

This chapter focuses on the recent reports on dicationic liquid crystals based on imidazolium motifs, both flexibly or rigidly linked bisimidazolium salts. The dicationic surfactants based on bisimidazolium salts show greatly enhanced properties, including significantly increased surface activity and lower cmc values (critical micelle concentration), higher thermal stability in comparison to traditional cationic or anionic surfactants or their corresponding monocationic counterparts. The chapter will review the structural factors that were found to influence the liquid crystalline properties of bisimidazolium salts: the type of spacer, the length of the flexible spacer between the imidazolium cations, the hydrophobic tails or various other mesogenic groups with more complex architecture, and the counterions. Finally, new perspectives in the design and targeted applications of the bisimidazolium based ILCs will be discussed.

Keywords: ionic liquid crystals, dicationic salts, imidazolium, bisimidazolium, surfactants

---

* Corresponding Author's Email: viorel.circu@chimie.unibuc.ro.

In: Imidazolium
Editor: Stephen A. Reyes
ISBN: 979-8-89113-425-6

Abbreviations

CMC	=	critical micelle concentration
Chol	=	cholesteryl unit
Col	=	columnar phase
Col_h	=	hexagonal columnar phase
Cub	=	cubic phase
DSC	=	differential scanning calorimetry
DS	=	dielectric spectroscopy
FTIR	=	Fourier-transform infrared spectroscopy
g	=	glassy state
ILs	=	ionic liquids
ILCs	=	ionic liquid crystals
LCs	=	liquid crystals
N	=	nematic
POM	=	polarizing optical microscopy
SmA	=	smectic A phase
SmB	=	smectic B phase
SmC	=	smectic C phase
SmT	=	smectic T phase
TGA	=	thermogravimetric analysis
XRD	=	X-ray diffraction

1. Introduction

Ionic liquid crystals (ILCs) are a distinct class of materials with unique behaviour resulting from the combination of liquid crystals and ionic liquids behaviour. Ionic liquids (ILs) are compounds completely composed of ions with melting points below 100°C (Lei et al., 2017). On the other hand, liquid crystals (LCs) are a state of matter that combines the properties of conventional isotropic liquids with those of crystalline solids. LCs present orientation order and lack (or have only a certain degree of) positional order and have found many applications in the construction of LCDs (liquid crystal displays) or various electro-optical devices and sensors. The topic of ILCs is continuously developing and many recent applications were added: battery materials, solar cells, electrochemical sensors, organic reaction media or electroluminescent switches that together increased its attractiveness.

By far, the ILCs based on imidazolium or pyridinium salts are very well documented (Kapernaum et al., 2022; Axenov and Laschat, 2011; Cîrcu, 2017); in fact, these salts are the most common ILCs studied for their exceptional characteristics such as low volatility, nonflammability, tunable polarity, high-ionic conductivity related to their ionic liquid nature in connection with their liquid crystalline properties. The related dimeric ionic liquids having a symmetric or unsymmetric structure with two head groups and two hydrocarbon chains, linked by a rigid or flexible spacer are called gemini surfactants (Menger and Littau, 1993). Gemini surfactants are an intensively studied class of amphiphilic molecules and many recent reviews were dedicated to this subject, including their synthesis and applications (Brycki et al., 2017, Guerrero-Hernández et al., 2022). Various combinations are possible for dissymmetric heterogemini surfactants with either two different, or the same, polar head groups and two different, or the same, hydrophobic groups. Compared with conventional monomeric surfactants, gemini surfactants demonstrate unique features in lowering the critical micelle concentration, good water solubility, reducing the surface tension, and even higher thermal stability. Consequently, various succesful applications were found as plasticizers (Kaur et al., 2022), emulsions, medical care, antimicrobial activity (Kowalczyk et al., 2020; Fatma et al., 2016; Brycki et al., 2021), enhanced oil recovery (Kamal, 2016), and steel corrosion inhibitor (Heakal and Elkholy, 2017) to name a few.

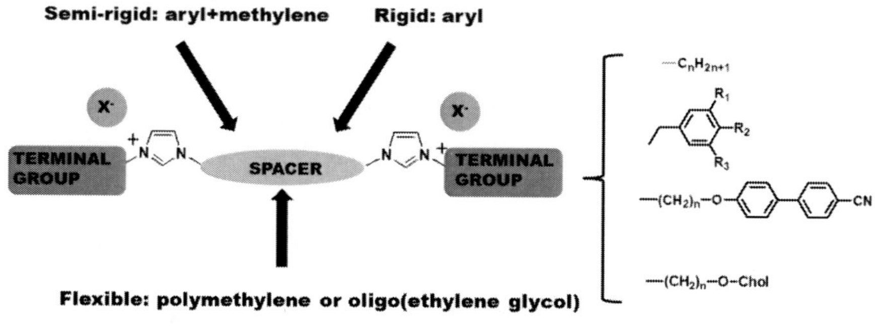

Figure 1. Schematic representation of the most common ILCs based on bisimidazolium salts.

Based on the simple scaffold of the gemini surfactants, with the right combination of spacer, counterion and hydrophobic tail, interesting potential liquid crystalline behaviour can be achieved. Moreover, the two hydrocarbons chains can be replaced by more elaborated mesogenic groups giving rise to ionic liquid crystals with tailored mesomorphic behaviour. This chapter will review the structural factors that were found to influence the liquid crystalline properties of bisimidazolium salts: the length of the flexible spacer between the imidazolium cations, the hydrophobic tails or the various mesogenic groups and the counterions. The intermolecular interactions such as hydrophobic interactions between the alkyl tails, ionic, dipole-dipole, anion–cation or hydrogen bonding play a major role in the stabilization of mesophases (liquid crystalline phases). Therefore, the mesogenic properties of bisimidazolium salts are very diverse, including calamitic materials (with N, SmA, SmC or SmT phases) or discotic materials with long-range columnar mesophases, related to mesogenic group employed and the nature of counterion (typical examples are: Cl^-, Br^-, NO_3^-, BF_4^-, PF_6^- or Tf_2N^-). It is worth mentioning that the most common phase seen for ILCs is the lamellar SmA phase. The stabilization of the lamellar phase is caused by the electrostatic interactions and ion-ion stacking specific to ILCs. The LC phases can be assigned by employing three characterization techniques: differential scanning calorimetry (DSC), polarizing optical microscopy (POM), and powder X-ray diffraction (XRD).

In this chapter, the discussion of LC properties will follow the classification of bisimidazolium-based ILCs which was made according to the nature of the spacer that connects the two imidazolium rings (flexible, semi-rigid and rigid) and of the mesogenic group used (either simple alkyl chains or more elaborated chemical units).

2. Bisimidazolium Salts with Flexible Spacers

The symmetric or unsymmetric dicationic bisimidazolium ionic liquid crystals with flexible spacers represent an interesting and versatile platform for molecular design. The studies reported so far were focused on systems with flexible polymethylene spacers. Less studied are the related systems having oligo(ethylene glycol) flexible spacers.

2.1. Symmetric Bisimidazolium Salts with Polymethylene Spacer and Terminal Alkyl Chains

The simplest bisimidazolium salts are those based on two imidazolium cations connected via polymethylene spacers and two terminal alkyl chains as presented in Figure 2.

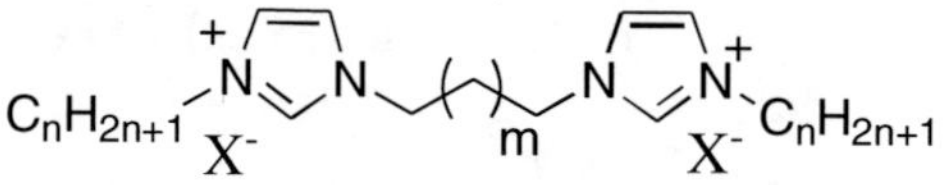

X = Br, **1-*m,n***	m= 4,6,8; n= 9,10,12,14,16
X = Cl, **2-*m,n***	m= 0 - 4, 6 - 8; n=6,8,10,12,14,16,18
X= BF_4, **3-*m,n***	m=0 - 4, 6 - 8; n= 4,6,7,8,10,11,12,14,16,18
X= PF_6, **4-*m,n***	m= 0 - 4, 6 - 8; n= 14,16,18
X=$AuCl_4$, **5-*m,n***	m=0 - 4, 6 - 8; n=18
X=$C_qH_{2q+1}OSO_3$, **6-*m,n,q***	m=4,8; n=14,16; q=10,12,14

Figure 2. Simple symmetric bisimidazolium salts with flexible polymethylene spacer.

2.1.1. Effect of the Spacer Length

Several general observations regarding the mesogenic behaviour of monocationic imidazolium salts can not be simply translated to the related bisimidazolium salts with flexible spacer and alkyl tails. Reports on the thermal behaviour of simple bisimidazolium salts evidenced the profound effect of the flexible spacer, e.g., short spacer in combination with sufficiently long hydrophobic tails display LC properties (smectic phases) (Huang et al., 2017; Ilincă, Pasuk, and Cîrcu, 2017; Robertson et al., 2013; Bara et al., 2010; Yang, Stappert and Mudring, 2014). Clearly, longer methylene spacers lead to lower clearing temperatures and prevent the stabilization of the LC phase. The decrease in the clearing temperatures for the bisimidazolium salts with long spacers is attributed to the lattice distortion of the rod-like molecular shape due to the existence of multiple conformations of the long polymethylene linkers (Huang et al., 2017).

Except for the report of Huang et al., (2017) on the chloride salts (**2-*m,n***), there are no other systematic studies on the effect of various counterions to show the maximum length of the spacer needed to preserve the mesophase. For example, salts with fixed anion and terminal alkyl chains show lower clearing temperatures up to a maximum number of carbon atoms in spacer after which the SmA phase is no longer stable: 8 carbon atoms in spacer for C16 and X=Cl (**2-*6,16***) and 9 carbon atoms in spacer for C18 and X=Cl (**2-*7,18***). Additional reports for other anions support these observations: for example, the bromide salts with C14 terminal alkyl tail (**1-*4,14***) show the formation of a SmA phase between 72°C and 230°C for the compound with 6 methylene groups in spacer while the analogue with 10 methylene groups (**1-*8,14***) in spacer melts straight to the isotropic phase around 100°C.

Table 1. Thermal data for bisimidazolium salts with flexible polymethylene spacers

Compound	X	Transition, T/°C
1-*4,4*	Br	Cr 50 Iso
1-*4,8*	Br	g -72 Iso
1-*4,10*	Br	Cr 52 SmA 58 Iso[a]
1-*4,10*	Br	Cr 72 Iso
1-*4,12*	Br	Cr 52 SmC 145 Iso[b]
1-*4,12*	Br	Cr 75 SmA 150 Iso[c]
1-*4,14*	Br	Cr 89 SmA >200 Iso
1-*4,14*	Br	Cr_1 56 Cr_2 72 SmA 230 Iso (dec.)
1-*6,12*	Br	Cr 114 Iso
1-*8,8*	Br	-[d]
1-*8,12*	Br	Cr 100 Iso
1-*8,14*	Br	Cr_1 61 Cr_2 101 Iso
1-*8,16*	Br	Cr 109 Iso
2-*2,6*	Cl	Liq.[e]
2-*2,8*	Cl	Cr <0 SmA 77 Iso
2-*2,10*	Cl	Cr 80 SmA 221 Iso
2-*2,12*	Cl	Cr 118 SmA 238 Iso
2-*2,14*	Cl	Cr 59 SmA 252 Iso
2-*2,16*	Cl	Cr 80 SmA 260 Iso
2-*2,18*	Cl	Cr 84 SmA 276 Iso
2-*0,18*	Cl	Cr_1 60 Cr_2 83 SmA 282 Iso
2-*1,18*	Cl	Cr 77 SmA 227 Iso
2-*3,18*	Cl	Cr 68 SmA 220 Iso
2-*4,18*	Cl	Cr 79 SmA 236 Iso
2-*6,18*	Cl	Cr 85 SmA 226 Iso
2-*7,18*	Cl	Cr 88 SmA 188 Iso

Compound	X	Transition, T/°C
2-8,18	Cl	Cr 93 Iso
2-0,16	Cl	Cr_1 76 Cr_2 104 SmA 282 Iso
2-1,16	Cl	Cr 48 SmA 227 Iso
2-3,16	Cl	Cr 71 SmA 252 Iso
2-4,16	Cl	Cr 73 SmA 246 Iso
2-6,16	Cl	Cr 80 SmA 226 Iso
2-7,16	Cl	Cr 83 Iso
3-0,4	BF_4	Cr 246 Iso[f]
3-0,6	BF_4	Cr_1 72 Cr_2 129 Iso
3-0,7	BF_4	Cr_1 28 Cr_2 75 Cr_3 158 Iso
3-0,8	BF_4	Cr_1 37 Cr_2 55 Cr_3 174 Iso
3-0,10	BF_4	Cr_1 34 Cr_2 56 Cr_3 85 Cr_4 191 Iso
3-0,11	BF_4	Cr_1 66 Cr_2 75 Cr_3 82 SmX 180 SmX 205 Iso
3-0,12	BF_4	Cr_1 42 Cr_2 76 Cr_3 90 SmX 187 SmX 254 Iso
3-0,18	BF_4	Cr_1 60 Cr_2 102 SmX 188 SmA 305 Iso
3-1,18	BF_4	Cr 98 SmA 305 Iso
3-2,18	BF_4	Cr 89 SmA 305 Iso
3-3,18	BF_4	Cr 59 SmA 253 Iso
3-4,18	BF_4	Cr 93 SmA 256 Iso
3-6,18	BF_4	Cr 61 SmA 190 Iso
3-7,18	BF_4	Cr 61 SmA 144 Iso
3-8,18	BF_4	Cr 79 SmA 91 Iso
3-8,16	BF_4	Cr 70 Iso
3-4,10	BF_4	Cr 70 Iso
3-4,12	BF_4	Cr 31 SmA 69 Iso
3-4,14	BF_4	Cr_1 57 Cr_2 86 SmA 167 Iso
3-6,12	BF_4	Cr 70 Iso
3-8,12	BF_4	Cr 55 Iso
4-0,18	PF_6	Cr_1 92 Cr_2 117 SmX 246 SmA 285 Iso
4-1,18	PF_6	Cr 103 SmB 259 SmA 310 Iso
4-2,18	PF_6	Cr 105 SmB 193 SmA 310 Iso
4-3,18	PF_6	Cr 84 SmA 310 Iso
4-4,18	PF_6	Cr 102 SmA 226 Iso
4-6,18	PF_6	Cr 93 SmA 164 Iso
4-7,18	PF_6	Cr 63 SmA 125 Iso
4-8,18	PF_6	Cr 62 SmA 84 Iso
4-4,14	PF_6	Cr 94 SmA 126 Iso
4-8,16	PF_6	Cr 57 Iso

Table 1. (Continued)

Compound	X	Transition, T/°C
5-0,18	$AuCl_4$	Cr_1 89 Cr_2 100 SmA 220 Iso
5-1,18	$AuCl_4$	Cr_1 95 Cr_2 101 SmA 220 Iso
5-2,18	$AuCl_4$	Cr 88 SmA 220 Iso
5-3,18	$AuCl_4$	Cr 79 SmA 220 Iso
5-4,18	$AuCl_4$	Cr 68 SmA 140 Iso
5-6,18	$AuCl_4$	Cr 83 Iso
5-7,18	$AuCl_4$	Cr 42 Iso
5-8,18	$AuCl_4$	Cr 80 Iso
6-4,14,10	$C_qH_{2q+1}OSO_3$	Cr 51 SmA 149 Iso
6-4,14,12	$C_qH_{2q+1}OSO_3$	Cr 65 SmA 148 Iso
6-4,14,14	$C_qH_{2q+1}OSO_3$	Cr 67 SmA 145 Iso
6-8,9,12	$C_qH_{2q+1}OSO_3$	Cr 38 SmA 48 Iso
6-8,14,10	$C_qH_{2q+1}OSO_3$	Cr_1 28 Cr_2 37 Cr_3 51 SmA 65 Iso
6-8,14,12	$C_qH_{2q+1}OSO_3$	Cr_1 48 Cr_2 60 Cr_3 70 SmA 85 Iso
6-8,14,14	$C_qH_{2q+1}OSO_3$	Cr_1 56 Cr_2 66 SmA 96 Iso
6-8,16,10	$C_qH_{2q+1}OSO_3$	Cr 39 SmA 81 Iso
6-8,16,12	$C_qH_{2q+1}OSO_3$	Cr 55 SmA 103 Iso
6-8,16,14	$C_qH_{2q+1}OSO_3$	Cr 64 SmA 114 Iso

[a] contradictory information reported by Yang, Stappert and Mudring (2014); SmA phase was assigned only on POM observations.

[b] contradictory information reported by Yang, Stappert and Mudring (2014); no experimental details provided for the assignment of the SmC phase.

[c] taken from Bara et al., (2010).

[d] no thermal data reported for this salt.

[e] in liquid state at room temperature.

[f] data taken from the first DSC heating run.

2.1.2. Effect of the Terminal Alkyl Chains

The bisimidazolium salts with the same spacer length and counterion exhibit an increase of the melting and clearing temperatures (the latter related to the higher stability of LC phase) for longer terminal chains. The trend is similar to the one found for monocationic imidazolium salts. For example, in the case of the bisimidazolium salts with PF_6 counterions and 10 methylene groups in the spacer, the elongation of the terminal alkyl tails from C16 to C18 represents the transition from a non-mesomorphic material (**4-*8,16***) to a SmA liquid crystal (**4-*8,18***) as the clearing point increases from 57°C to 84°C. The same observation can be easily made for the bisimidazolium salts with alkyl sulfates as counterions. As an example, the clearing temperature of **6-*8,14,10*** is 65°C while for **6-*8,16,10*** is 81°C (Ilincă, Pasuk, and Cîrcu, 2017).

2.1.3. Effect of the Counterion

In general, the LC properties seen for short spacer salts resemble the general trend found for ILCs, where clearing points show a decreasing tendency with an increase in the size of anions and a lower ability to form hydrogen bonding with the imidazolium cations. Usually, the highest clearing temperatures are observed for chloride and bromide bisimidazolium salts due to the strength of hydrogen bonding between the imidazolium cations and the halide anions. For example, the SmA phase of the bromide salt **1-*4,12*** is stable up to 150°C, significantly higher compared to the tetrafluoroborate salt **3-*4,12*** for which the transition from SmA to isotropic state occurs at 69°C. The isotropization temperatures decrease from 230°C with decomposition (X=Br, **1-*4,14***) to 167°C (X=BF_4, **3-*4,14***) and to 126°C (X=PF_6, **4-*4,14***) when the halide anion was replaced with larger size ions. For bisimidazolium salts with $AuCl_4$ counterions and shorter spacer, the stability of the mesophase is drastically reduced, in the majority of the examples with almost 100°C (see **2-*4,18***, X=Cl, 236°C compared to **5-*4,18***, X=$AuCl_4$, 140°C) (Huang et al., 2017). It is worth mentioning here that the tetrachloroaurate(III) bisimidazolium salts (**5-*m,n***) were successfully used as gold precursor sources as well as capping/stabilizing agents for gold nanoparticles (AuNPs). Interestingly, no significant effect of the number of carbon atoms contained in the alkyl sulfate anion (**q**) on the clearing temperatures of the related bisimidazolium salts with shorter spacer (**6-*m,n,q***) was observed (145°C for **6-*4,14,14***, 148°C for **6-*4,14,12*** and 149°C for **6-*4,14,10***, respectively). The size of alkyl sulfate anions has a much more pronounced effect on the clearing temperature for the two series with longer spacer, **6-*8,14,q*** and **6-*8,16,q***. The increase of the number of carbon atoms of alkyl sulfate anion yields higher isotropization temperatures, for example, 81°C for **6-*8,16,10*** compared to 114°C for **6-*8,16,14*** (Ilincă, Pasuk, and Cîrcu, 2017).

2.2. Unsymmetric Bisimidazolium Salts with Polymethylene Spacer

The unsymmetric hexyl-bridged bisimidazolium salts with only one alkyl chain have been reported by Robertson et al., (2013) (Figure 3). These bisimidazolium salts with various counterions (X=Br, BF_4 or Tf_2N) can form thermotropic bicontinuous cubic liquid crystal phases in neat form and/or lyotropic bicontinuous cubic phases with several non-aqueous solvents or water (Table 2). The related salts **9-*14*** and **9-*16*** show no mesogenic properties.

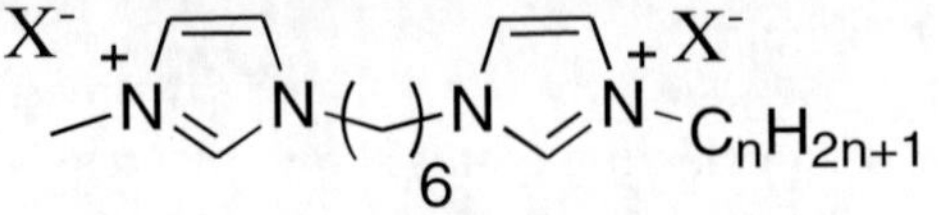

X = Br (**7-*n***), BF_4 (**8-*n***), Tf_2N (**9-*n***)
n=14, 16, 18, 20

Figure 3. Usymmetric bisimidazolium salts with only one terminal alkyl chain.

Table 2. Thermal data for unsymmetric bisimidazolium salts

Compound	X	Transition, T/°C
7-*14*	Br	Cr 18 SmX[b] 184 dec.
7-*16*	Br	Cr 40 LC[a] 63 Cub_{bi} 189 dec.
7-*18*	Br	Cr 57 SmX[b] 192 dec.
7-*20*	Br	Cr 66 SmX[b] 225 dec.
8-*14*	BF_4	Cr 19 SmX[b] 146 Iso
8-*16*	BF_4	Cr 40 SmX[b] 180 Iso
8-*18*	BF_4	Cr (57 LC[a])[c] 58 Cub_{bi} 220 dec.
8-*20*	BF_4	Cr (64 LC[a])[c] 67 Cub_{bi} 210 dec.
9-*18*	Tf_2N	Cr (39 SmX[b])[c] 47 Iso
9-*20*	Tf_2N	Cr (67 SmX[b])[c] 52 SmX 71 Iso

[a] LC - unidentified mesophase.
[b] SmX – unidentified smectic phase.
[c] monotropic transition.

These salts display amphotropic character, meaning that they show both thermotropic and lyotropic behaviour. A lyotropic phase type Q in various solvents such as water, glycerol, and the ionic liquid (IL) ethylmethylimidazolium tetrafluoroborate ($EMIMBF_4$) was evidenced by SAXS measurements for compounds with bromide and tetrafluoroborate anions (Robertson et al., 2013).

2.3. Bisimidazolium Salts with Oligo(ethylene glycol) Spacers

Symmetric, dicationic imidazolium ionic liquid crystals with poly(ethylene glycol) spacer were reported in two separate papers by Bara et al., (2010) and by Yang, Stappert and Mudring (2014). By comparison with the related salts with methylene spacer, the bisimidazolium salts with poly(ethylene glycol) spacer were found to have higher stability of the LC phase as well as richer mesomorphism. Compounds **11-*1,12*** and **12-*1,12*** with tetrafluoroborate anion and perchlorate anion respectively display a rare higher-order smectic T phase, which changed to a smectic A phase during heating (see Table 3).

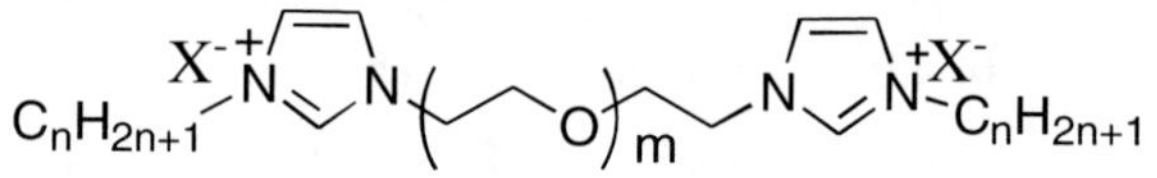

C_nH_{2n+1} X^- N+ N (O)m N N+ X^- C_nH_{2n+1}

X =Br (**10-*m,n***), BF_4 (**11-*m,n***), ClO_4 (**12-*m,n***), PF_6 (**13-*m,n***), NTf_2 (**14-*m,n***)
m = 1-3; n = 4,8,10,12,14

Figure 4. Flexible bisimidazolium salts with oligo(ethylene glycol) spacers.

Table 3. Thermal data for bisimidazolium salts with oligo(ethylene glycol) spacers

Compound	X	Transition, T/°C
10-*1,4*	Br	g -23 Iso
10-*1,8*	Br	Cr 93 Iso
10-*1,10*	Br	Cr 5 SmC 102 SmA 128 Iso
10-*1,12*	Br	Cr_1 4 Cr_2 46 SmC 96 SmA 200 Iso
11-*1,10*	BF_4	Cr 22 SmA 27 Iso
11-*1,12*	BF_4	Cr 22 SmA 148 Iso[a]
11-*1,12*	BF_4	Cr 6 SmT 20 SmA 140 Iso[b]
11-*1,14*	BF_4	Cr 42 SmA 200 Iso
11-*2,12*	BF_4	Cr 29 SmA 50 Iso
11-*3,12*	BF_4	Cr 22 SmA 37 Iso
12-*1,12*	ClO_4	Cr -1 SmT 28 SmA 126 Iso
13-*1,12*	PF_6	Cr 50 SmA 147 Iso
14-*1,12*	Tf_2N	Cr (Cr_1 -22 Cr_2 -17 SmA -13 SmX 2 Cr_3 26)[c] 27.5 Iso

[a] taken from Bara et al., (2010).
[b] taken from Yang, Stappert and Mudring (2014).
[c] monotropic transitions.

On the other hand, the alkyl ether units as polar groups have been shown to exhibit attractive hydrogen bonding interactions with imidazolium cations that aid to stabilize the lamellar organization of the gemini amphiphiles.

The anion type has a certain influence on the transition temperatures as similarly observed for salts with methylene spacers and almost no major impact on the thermotropic LC phase formation for analogous compounds. For example, the clearing temperatures in the series **10-*1,12*** (X=Br), **11-*1,12*** (X=BF_4), **12-*1,12*** (X=ClO_4), **13-*1,12*** (X=PF_6) and **14-*1,12*** (X=Tf_2N) show a descending trend as follows: 200°C, 148°C, 126°C, 147°C, and 27.5°C, respectively (Table 3). Indeed, this is the expected trend when the melting and clearing temperatures are strongly related to the size of counterions: lower transition temperatures for the larger counterion.

2.4. Symmetric Bismidazolium Salts with Polymethylene Spacer and Other Terminal Mesogenic Groups

Many studies have been devoted to bisimidazolium salts with flexible polymethylene spacer and more elaborated mesogenic groups: 4-, 3,4- and 3,4,5-substituted benzyl moieties, cyanobiphenyl or cholesteryl units.

2.4.1. Bisimidazolium Salts with Substituted Benzyl Moieties

Bisimidazolium salts with short propylene spacer and 4-alkyloxy (**15-*n*** and **16-*n***) or 3,4-dialkyloxy-substituted (**17-*n***) benzyl moieties as terminal groups were reported by Baron et al., (2016) (Figure 5). All these salts show enantiotropic SmA phase (Table 4) with no significant differences between the mesogenic character of the dicatenar and the tetracatenar compounds. The bisimidazolium salts were used as precursors for the preparation of N-heterocyclic dicarbene gold(I) complexes. These gold-carbene products were obtained by deprotonation in dimethylformamide of the corresponding bromide salts with a mild base (NaOAc) in the presence of $AuCl(SMe_2)$ as the gold precursor. Only the gold-carbene complexes based on the tetracatenar systems preserved mesomorphism while the related complexes derived from the dicatenar salts lost mesogenicity and simply presented several crystalline phases.

The symmetric dicationic hexacatenar imidazolium ionic liquids with 3,4,5-trisalkyloxy substituted benzyl moieties and chloride or tetrafluoroborate anions (**18-*n,m***, X= Cl and **19-*n,m***, X=BF_4) were found to exhibit Col_h mesophases whereas the systems with the bulkier bis(trifluoromethylsulfonyl)imide (**20-*n,m***, X=Tf_2N) anion melted directly to an isotropic liquid (Gao, Slattery and Bruce, 2011).

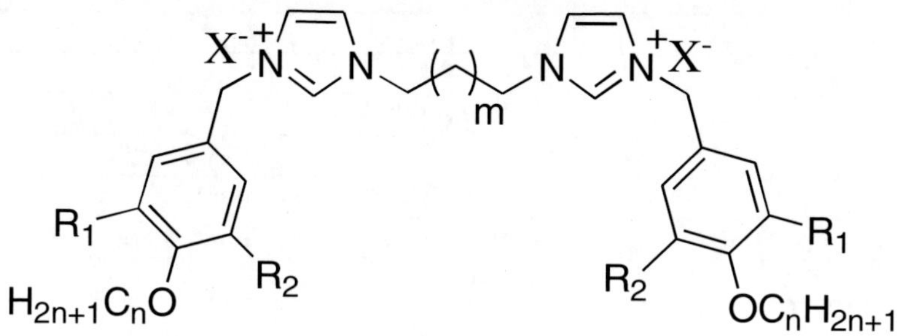

$R_1 = R_2 = H$
X = Br (**15-*n***), PF_6 (**16-*n***)
n = 8, 10, 14, 16 (a-e) m = 1

$R_1 = OC_{12}H_{25}$ $R_2 = H$
X=Br (**17-*n***)
n=12 m=1

$R_1 = R_2 = OC_nH_{2n+1}$
X = Cl (**18-*n,m***), BF_4 (**19-*n,m***), Tf_2N (**20-*n,m***)
n = 8,12 (a,b) m = 2, 4, 6

Figure 5. Flexible bisimidazolium salts with substituted benzyl moieties as mesogenic groups.

Importantly, the hexagonal columnar mesophase of these salts is stable on an extremely large interval of more than 200°C. Some of these salts show also the Col_h phase stable at room temperature. The LC properties of these salts were influenced by the spacer length, terminal alkyl length and the nature of counterions. Thus, the largest mesophase ranges were found for bisimidazolium salts with longer alkyl chain lengths and short spacer lengths.

Table 4. Thermal data for bisimidazolium salts with benzyl mesogenic groups

Compound	X	Transition, T/oC
15-*8*	Br	(Cr+SmA) 80 SmA 190 dec.
15-*10*	Br	(Cr+SmA) 80 SmA 200 dec.
15-*16*	Br	Cr_1 46 Cr_2 70 Cr_3 111 SmA 200 dec.
16-*12*	PF_6	Cr_1 100 Cr_2 130 SmA 190 dec.
17-*12*	Br	Cr_1 46 Cr_2 70 Cr_3 110 SmA 200 dec.
18-*8,2*	Cl	Cr 46 Col_h 241.4 Iso
18-*8,4*	Cl	Cr_1 62 Cr_2 81.4 Col_h 237 Iso
18-*8,6*	Cl	Cr_1 -20.6 Cr_2 34 Col_h 184 Iso
18-*12,2*	Cl	Cr_1 49 Cr_2 72 Col_h 213 Iso
18-*12,4*	Cl	Cr_1 36 Cr_2 100.5 Col_h 221 Iso
18-*12,6*	Cl	Cr 36.7 Col_h 215.8 Iso
19-*8,2*	BF_4	Cr 34 Col_h 238 Iso
19-*8,4*	BF_4	Cr 41 Col_h 185.0 Iso
19-*8,6*	BF_4	Cr 98 Col_h 135 Iso
19-*12,2*	BF_4	Cr 4.0 Col_h 256 Iso
19-*12,4*	BF_4	Cr 6 Col_h 246 Iso
19-*12,6*	BF_4	Cr_1 36 Cr_2 51 Col_h 213 Iso

2.4.2. Bisimidazolium Salts with Other Mesogenic Groups

A series of bisimidazolium salts with various mesogenic groups (cyanobiphenyl or cholesteryl attached via a flexible methylene chain) and their silver carbene complexes with Br^- anion has been designed and prepared by us (Pană et. al., 2014) (Figure 6). The metal carbenes were prepared by the reaction of bisimidazolium bromide salts with Ag_2O in dichloromethane. The bisimidazolium salt having cholesteryl groups (**24-*8***) shows higher transition temperatures than the compounds possessing cyanobiphenyl mesogenic groups, but the thermal stability of SmA phase is limited by the decomposition that occurs during the transition to isotropic state at 135°C.

It is very interesting to note that a short-range nematic phase (only 2°C) was observed for the tetracatenar bisimidazolium salt containing four cyanobiphenyl mesogenic groups on each side, which is very uncommon for ionic liquid crystals, in particular for imidazolium-based ILCs (Pană et al., 2014). Moreover, all these bisimidazolium salts and the corresponding silver carbene complexes with cyanobiphenyl mesogenic groups are intense blue emitters, both in solution and solid state, with high quantum yields in the 60-

77% range. The effect of titanium oxide nanoparticles on the dielectric properties and ionic conductivity of the bisimidazolium salt with dodecyl sulfate anion (**22-*4,4***), which display a SmA phase stable at room temperature, was studied by Ganea et al., TiO_2 nanoparticles were mixed with **22-*4,4*** in concentrations of 0.1%, 0.2%, 1% and 2%. An increase of the clearing points was observed for the higher concentration of TiO_2 nanoparticles, from 35°C for pure ILC to 39°C for ILC doped with 2% TiO_2 accompanied by the sharpening of the SmA-Iso transition, as expected (Ganea, Cîrcu and Manaila-Maximean, 2020). It was observed that at a constant frequency, the ionic conductivity increased with the temperature and dopant concentration (around 10^{-6}S·cm^{-1} at 45°C). The first examples of discotic liquid crystalline gemini surfactants (**25-*m,n***) were reported by Kumar and Gupta (2010). These salts exhibit hexagonal columnar mesophase over a wide temperature range which is influenced by the spacer length and display ionic conductivity in the range of 10^{-6} to 10^{-5} S·cm^{-1}.

Table 5. Thermal data for bisimidazolium salts with cyanobiphenyl or cholesteryl mesogenic groups

Compound	X	Transition, T/oC
21-*8,8*	Br	Cr 34 SmA 55 Iso
22-*4,4*	$C_{12}H_{25}OSO_3$	Cr 5 SmA 35 Iso
23-*8,8*	Br	g 51 SmA 79 N 81 Iso
24-*8*	Br	g 39 SmA 135 dec
25-*4,6*	Br	Cr 48 Col_h 180 Iso
25-*5,12*	Br	Cr 60 Col_h 121 Iso
25-*8,8*	Br	Cr 59 Col_h 87 Iso
25-*8,9*	Br	Cr 59 Col_h 86 Iso
25-*8,10*	Br	Cr 58 Iso
25-*8,12*	Br	Cr 58 Iso

3. Bisimidazolium Salts with Semi-Rigid Spacer

Various semi-rigid spacers were used to connect the two imidazolium organic cations. They are formed by joining an aromatic rigid core and at least one methylene group which is directly attached to the nitrogen atom of the imidazolium ring. A selection of the most interesting systems is presented in Figure 7.

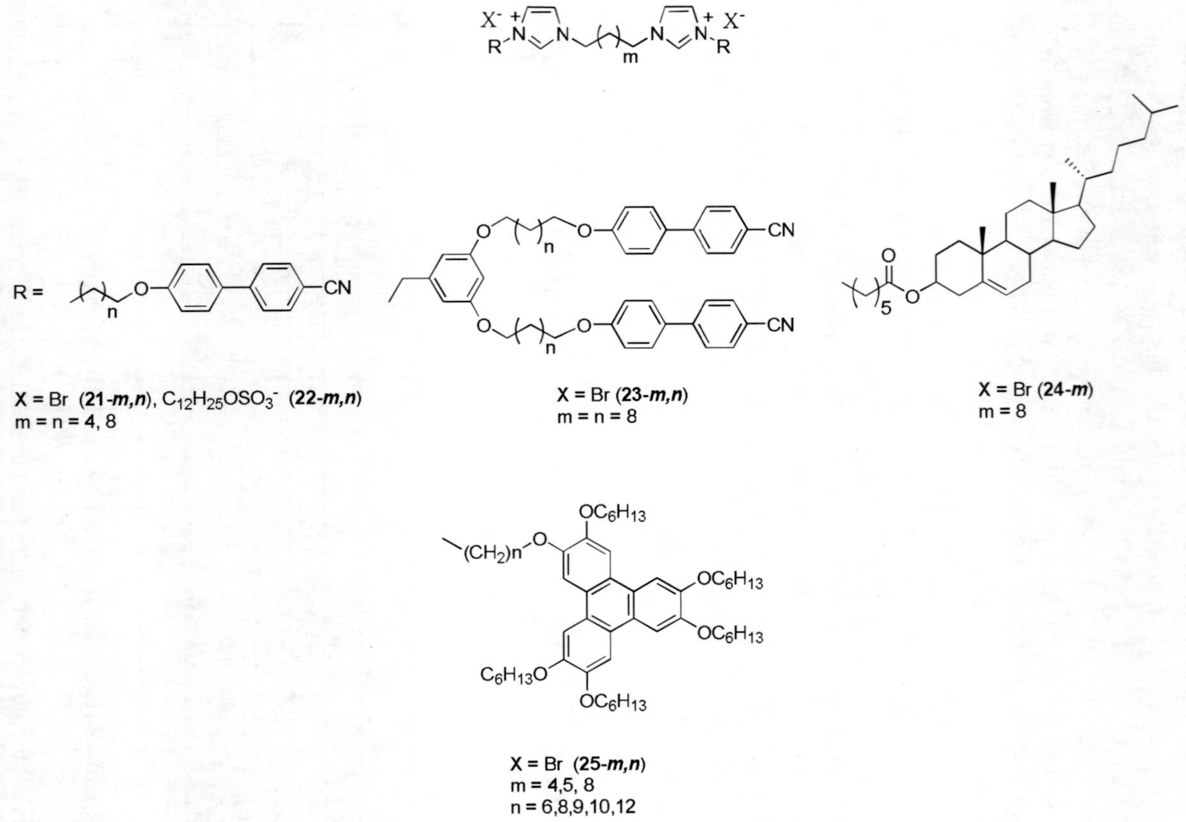

X = Br (**21-*m,n***), $C_{12}H_{25}OSO_3^-$ (**22-*m,n***)
m = n = 4, 8

X = Br (**23-*m,n***)
m = n = 8

X = Br (**24-*m***)
m = 8

X = Br (**25-*m,n***)
m = 4,5, 8
n = 6,8,9,10,12

Figure 6. Flexible bisimidazolium salts with cyanobiphenyl or cholesteryl units as mesogenic groups.

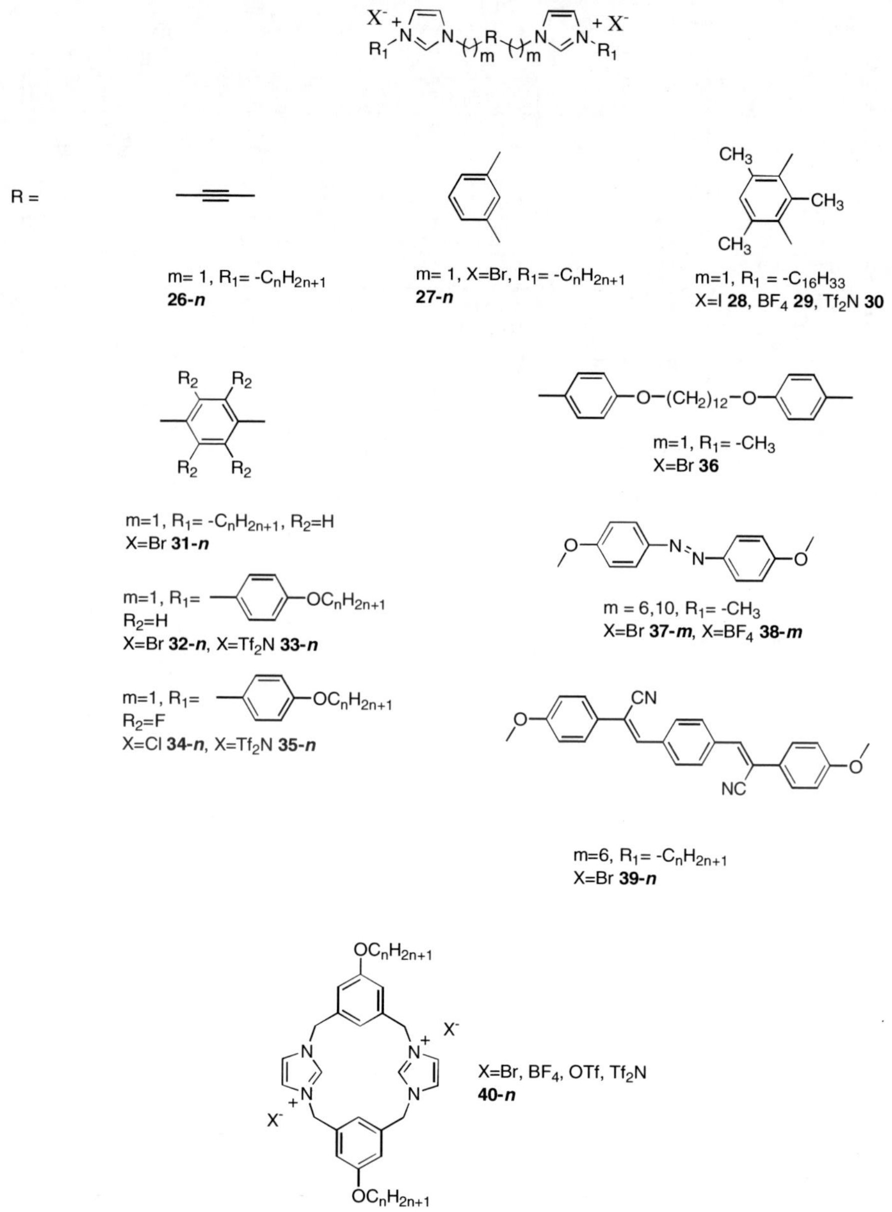

R =

m= 1, R_1= -C_nH_{2n+1}
26-*n*

m= 1, X=Br, R_1= -C_nH_{2n+1}
27-*n*

m=1, R_1 = -$C_{16}H_{33}$
X=I **28**, BF_4 **29**, Tf_2N **30**

m=1, R_1= -C_nH_{2n+1}, R_2=H
X=Br **31-*n***

m=1, R_1= OC_nH_{2n+1}
R_2=H
X=Br **32-*n***, X=Tf_2N **33-*n***

m=1, R_1= OC_nH_{2n+1}
R_2=F
X=Cl **34-*n***, X=Tf_2N **35-*n***

m=1, R_1= -CH_3
X=Br **36**

m = 6,10, R_1= -CH_3
X=Br **37-*m***, X=BF_4 **38-*m***

m=6, R_1= -C_nH_{2n+1}
X=Br **39-*n***

X=Br, BF_4, OTf, Tf_2N
40-*n*

Figure 7. Examples of bisimidazolium salts with semi-rigid spacers.

Table 6. Thermal data for bisimidazolium salts with semi-rigid spacer

Compound	X	Transition, T/°C
26-*4*	Br	Cr 133 Iso
26-*8*	Br	Cr 75 Iso
26-*10*	Br	Cr 80 SmC dec.
26-*12*	Br	Cr_1 58 Cr_2 81 SmC dec.
27-*10*	Br	Sm[a] 57 Iso
27-*18*	Br	Cr 61 Sm[a] 250 Iso
28	I	Cr 33 SmA 157 Iso
29	BF_4	Cr 43 SmA 150 Iso
30	Tf_2N	Cr 63 Iso
31-*4*	Br	Cr_1 30 Cr_2 71 Iso
31-*8*	Br	Cr 87 Iso
31-*10*	Br	Cr_1 35 Cr_2 62 Iso
31-*12*	Br	Cr_1 33 Cr_2 84 SmC 162 SmA 234 Iso (dec.)
32-*12*	Br	Cr 106 SmA 299 Iso (dec.)
33-*12*	Tf_2N	Cr 85 Iso
34-*8*	Cl	Cr_1 36 Cr_2 56 Iso
34-*10*	Cl	Cr 50 Iso
34-*12*	Cl	Cr 60 Iso
35-*8*	Tf_2N	Cr 64 Iso
35-*10*	Tf_2N	Cr 74 Iso
35-*12*	Tf_2N	Cr_1 63 Cr_2 (SmA 79)[b] 82 Iso
36	Br	Cr 82 SmA 127 Iso
37-*6*	Br	Cr 154 Iso
37-*10*	Br	Cr_1 77 Cr_2 156 SmC 169 Iso
38-*6*	BF_4	Cr 143 Iso
38-*10*	BF_4	Cr_1 77 Cr_2 98 Cr_3 155 Iso
39-*1*	Br	Cr 91 Sm[a] 205 Iso
39-*4*	Br	Cr 84 Sm[a] 183 Iso
39-*6*	Br	Cr 72 Sm[a] 162 Iso

[a] type of the smectic phase was not assigned.
[b] monotropic phase.

Bisimidazolium salts with butyne spacer (**26-*n***) are able to form mesophases for longer alkyl chains (10 and 12) (Yang, Stappert and Mudring, 2014). These salts have poor thermal stabilities that could be associated with the breaking of the rigid triple C–C bond during the heating runs, leading to the decomposition of the compounds at very low temperatures, around 80°C for longer chain analogues. Bisimidazolium salts with various counterions (**28**, X=I; **29**, X=BF_4, and **30**, X=Tf_2N) bearing terminal hexadecyl chains and a bridging mesitylene moiety were reported by Trilla et al., (2008). The diffusion nuclear magnetic resonance (NMR) experiments showed that bisimidazolium iodide salt **28** can form self-aggregates in solution. The larger size Tf_2N anion caused the lack of mesomophism for compound **30**. The bisimidazolium salts with two cationic rings linked by a 1,3-dimethylenebenzene spacer **27-*n*** and related macrocyclic derivatives **40-*n*** were developed by Casal-Dujat et al., (2012). The latter compounds display a higher melting and clearing point than the related counterparts which could be explained by the lower conformational flexibility of the cyclophanes. Salt with a shorter alkyl chain **27-*10*** exists in LC state at room temperature.

The bisimidazolium salts with a linker benzene ring with 1,4- connecting positions (**31-*n***) have been developed by Yang, Stappert and Mudring (2014). The melting points of these salts are relatively high due to the π–π interactions caused by the rigid aromatic benzene spacer. There is a need of longer alkyl chains to stabilize the LC phase of these salts as the analogues with shorter chains (4, 8 or 10) do not display LC properties. The analogues with dodecyl terminal chains showed a very large (~150°C) liquid crystalline temperature range and a rare SmC phase.

The related bisimidazolium salts tethered with 4-alkoxyphenyl groups (**32-*12*** and **33-*12***) together with a series of new dicationic bisimidazolium salts bearing fluoro substituents on the linker benzene-ring positions (**34-*n*** and **35-*n***) have been synthesized and characterized by Li, Bruce and Shreeve, (2009). This investigation revealed that the fluorination of the aryl linker does not improve the mesogenic properties. Only compound **35-*12*** showed a monotropic SmA phase on cooling from the isotropic phase.

A more elongated semi-rigid linker comprising two benzene rings connected via a dodecamethylene spacer was used to prepare new bolaform bisimidazolium ILs and ILCs (**36**). Al Abbas et al., (2017) investigated the influence of the form and charge of the counteranion, and the bromide anion was replaced with a series of cyanometalates: $[Ag(CN)_2]^-$, $[Ni(CN)_4]^{2-}$, $[Pt(CN)_4]^{2-}$, $[Co(CN)_6]^{3-}$, $[Fe(CN)_6]^{3-}$. The new resulting bisimidazolium metallomesogens derived from **36** exhibited lamellar organization (SmA) except for those containing the square-planar $[Ni(CN)_4]^{2-}$ and $[Pt(CN)_4]^{2-}$ species. The authors did not investigate other properties related to the presence of metal centers. Zhang et al., (2008) reported the bisimidazolium salts with an azobenzene-containing linker (**37-*m*** and **38-*m***). It is well known that the azobenzene units were intensively used as a mesogenic group in liquid crystals, including ILCs. The azobenzene unit is part of the structure of many ILCs with rare SmC phase. Salt **37-*10*** with longer semi-rigid linker show a SmC phase stable at high temperatures from 156°C to 169°C. The related products do not show any LC properties reflecting the effect of the much stronger interactions between the anion and the imidazolium cations needed to stabilize the mesophase. The authors have found no influence on the arrangement and configuration of the azobenzene groups in solution based on UV-VIS spectroscopy results.

Recently, Wu et al., (2022) developed the bisimidazolium bromide salts with dicyanodivinylbenzene linker (**39-*n***) intending to report the first gemini-type fluorescence ILCs. But, there are several reports dealing with the emission properties of such dicationic imidazolium salts. For example, salts **21-*m,n*** and **23-*m,n*** are good emitters in solid state and in solution with high quantum yields, in the 60-77% range (Pană et. al., 2014). The LC properties of salts **39-*n*** depend on the polymethylene spacer used to connect the imidazolium cation and the dicyanodivinylbenzene unit. All of them display an enantiotropic smectic phase on a long thermal range with higher stability for shorter polymethylene spacers. Interestingly, these salts were able to form stable gels in DMSO-toluene (1:1) system. Only the long spacer analogue **39-*6*** showed a rectangular columnar stacking in gel, as evidenced by XRD and FTIR measurements (Wu et al., 2022).

4. Bisimidazolium Salts with Rigid Spacer

There are only a few reports on bisimidazolium salts with rigid aromatic spacer. Their chemical structure is presented in Figure 8 and the corresponding thermal behavior is given in Table 7.

The mesogenic properties of the bisimidazolium salts with rigid aryl spacers are strongly affected by the counterion. While the bromide salts **41-*Ph,12*** and **41-*Ph,16*** show no liquid crystalline properties, the related salts with OTf counterions display a rare SmC phase with a much larger temperature range for the longer alkyl tails (n = 16) (Noujeim et al., 2012). When the counterion was replaced with Tf_2N, the new SmT phase was seen at lower temperatures.

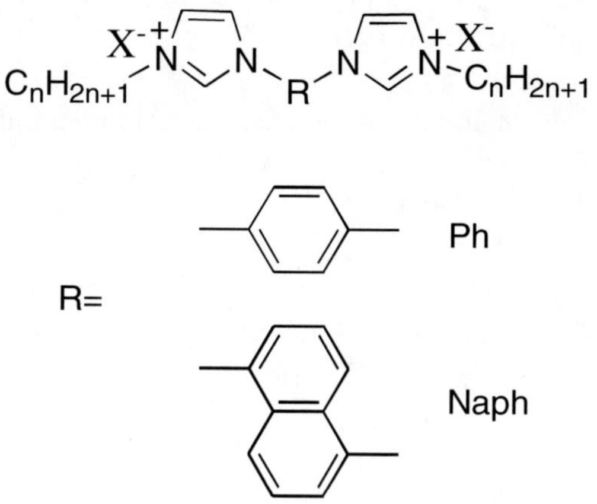

X =Br (**41-*n***), BF_4 (**42-*n***), PF_6 (**43-*n***), OTf (**44-*n***), Tf_2N (**45-*n***)
n = 12,16

Figure 8. Bisimidazolium salts with rigid spacer.

The same highly organized SmT phase was also observed for the related salts with naphthalene connecting unit (**45-*Naph,12*** and **45-*Naph,16***) but on narrower thermal ranges and lower isotropization temperatures. Several reports indicated that the naphthalene core leads to lower transition temperatures of various LCs (Lehmann, Jahr and Gutmann, 2008). These latter examples of ILCs have been successfully used as organized media for the intramolecular Diels Alder reaction with high yields and without using high dilution conditions (Doa and Schmitzer, 2015). The remaining salts with naphthalene spacer and various counterions (Br, BF_4, PF_6 and OTf) did not show liquid crystalline phases.

Table 7. Thermal data for bisimidazolium salts with rigid spacer

Compound	R	X	Transition, T/oC
44-*Ph,12*	Ph	OTf	Cr_1 148 Cr_2 169 SmC 184 Iso
44-*Ph,16*	Ph	OTf	Cr_1 141 Cr_2 149 SmC 250 dec.
45-*Ph,12*	Ph	Tf_2N	Cr 55 SmT 136 Iso
45-*Ph,16*	Ph	Tf_2N	Cr 75 SmT 172 dec.
45-*Naph,12*	Naph	Tf_2N	Cr_1 87 Cr_2 93 SmT 123 Iso
45-*Naph,16*	Naph	Tf_2N	Cr 94 SmT 166 Iso

5. Miscellaneous Bisimidazolium Salts

An interesting example of thermotropic ionic liquid crystalline lithium salt was proposed by Yuan et al., (2019) for potential application in Li-ion batteries. Based on DSC and POM observations, the authors suggested the existence of a mesophase from 72°C down to 39°C without its assignment. It was found that, in solid state, salt **46** has higher electrochemical stability and ionic conductivity when compared to traditional lithium salts.

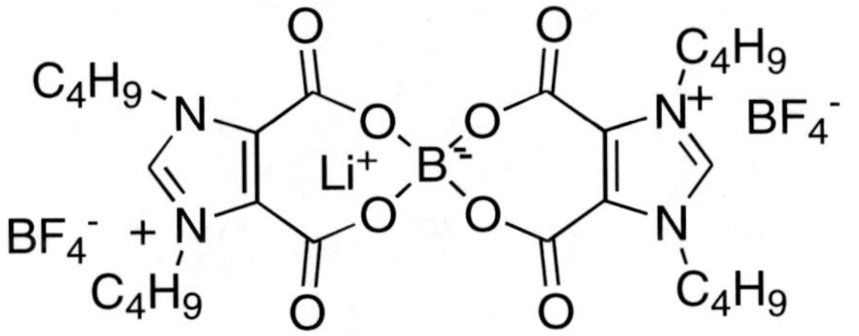

46

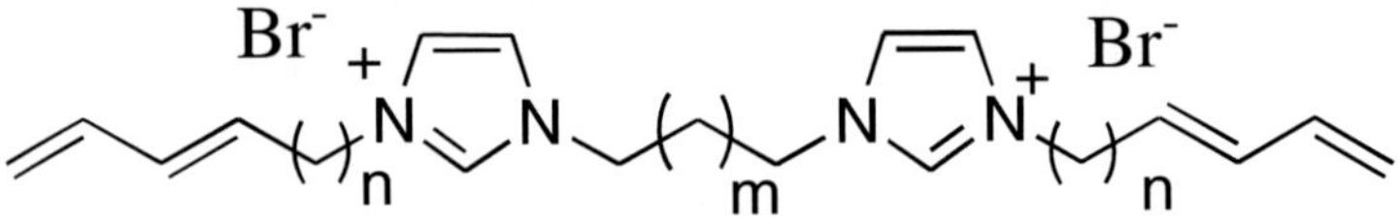

47-m,n

m = 2-7
n = 10, 14

Figure 9. Other bisimidazolium salts.

A series of (alkyldiene-imidazolium bromide) gemini salts (**47-*m,n***) with flexible polymethylene spacer and polymerizable diene tails were prepared and tested for their ability to form a bicontinous cubic (Q) lyotropic liquid crystals phase with glycerol or water. The formation of Q-phase was evidenced only for homologues that have shorter spacers (five or six carbon atoms, m = 3 or 4), with no influence of the diene tails (Li et al., 2021; Dwulet et al., 2019).

Conclusion

Since the report by Bara et al., in 2010, the field of ILCs based on bisimidazolium platform has tremendously developed toward more complex molecular architecture incorporated in both the spacer and the terminal mesogenic groups. The intensive research in this field that has been reviewed in this chapter helped to elucidate the important structure-properties relationship for bisimidazolium ILCs. Further investigations should take into account the complexity of the structural parameters not only for preserving or increasing the stability of the mesophase, but to add and develop new and significant functionalities such as ionic conductivity, luminescence, the ability to catalyze specific organic reactions carried out in organized media, electro-optical effects, etc. There are also opportunities to develop commercial applications based on ILCs that can ensure the growth of the field in the next future.

References

Al Abbas, A., Heinrich, B., L'Her, M., Couzigné, E., Welter, R. and Douce, L. (2017). Bolaamphiphilic Liquid Crystals Based on Bis-imidazolium Cations, *New Journal of Chemistry*, 41, 2604-2613.

Axenov, K. V. and Laschat, S. (2011). Thermotropic Ionic Liquid Crystals, *Materials*, 4, 206–259.

Bara, J. E., Hatakeyama, E. S., Wiesenauer, B. R., Zeng, X., Noble, R. D. and Gin, D. L. (2010). Thermotropic liquid crystal behaviour of gemini imidazolium-based ionic amphiphiles, *Liquid Crystals*, 37, 1587–1599.

Baron, M., Bellemin-Laponnaz, S., Tubaro, C., Heinrich, B., Basato, M. and Accorsi, G. (2016). Synthesis and thermotropic behaviour of bis(imidazolium) salts bearing long chain alkyl-substituents and of the corresponding dinuclear gold carbene complexes, *Journal of Organometallic Chemistry*, 801, 60-67.

Brycki, B. E., Kowalczyk, I. H., Szulc, A., Kaczerewska, O. O. and Pakiet, M. (2017). Multifunctional Gemini Surfactants: Structure, Synthesis, Properties and Applications, in *Application and Characterization of Surfactants*. London, United Kingdom: IntechOpen. Available: https://www.intechopen.com/chapters/55368 doi: 10.5772/intechopen.68755.

Brycki, B. E., Szulc, A., Kowalczyk, I., Koziróg, A. and Sobolewska, E. (2021). Antimicrobial Activity of Gemini Surfactants with Ether Group in the Spacer Part, *Molecules*, 26, 5759.

Casal-Dujat, L., Penon, O., Rodriguez-Abreu, C., Solans, C. and Perez-Garcia, L. (2012). Macrocyclic ionic liquid crystals, *New Journal of Chemistry*, 36, 558–561.

Cîrcu, V. (2017). Ionic liquid crystals based on pyridinium salts, in *Progress and Developments in Ionic Liquids* (Ed.: S. Handy), IntechOpen, 285–311.

Doa, T. D. and Schmitzer, A. R. (2015). Intramolecular Diels Alder reactions in highly organized imidazolium salts-based ionic liquid crystals, *RSC Advances*, 5, 635-639.

Dwulet, G. E., Dischinger, S. M., McGrath, M. J., Basalla, A. J., Malecha, J. J., Noble, R. D. and Gin, D. L. (2019). Breathable, Polydopamine-Coated Nanoporous Membranes That Selectively Reject Nerve and Blister Agent Simulant Vapors, *Industrial and Engineering Chemistry Research*, 58, 47, 21890–21893.

Fatma, N., Panda, M. and Beg, M. (2016). Ester-bonded cationic gemini surfactants: Assessment of their cytotoxicity and antimicrobial activity, *Journal of Molecular Liquids*, 222, 390–394.

Ganea, C. P., Cîrcu, V. and Manaila-Maximean, D. (2020). Effect of titanium oxide nanoparticles on the dielectric properties and ionic conductivity of a new smectic bis-imidazolium salt with dodecyl sulfate anion and cyanobiphenyl mesogenic groups, *Journal of Molecular Liquids*, 317, 113939.

Gao, Y., Slattery, J. M. and Bruce, D. W. (2011). Columnar thermotropic mesophases formed by dimeric liquid-crystalline ionic liquids exhibiting large mesophase ranges, *New Journal of Chemistry*, 35, 2910–2918.

Guerrero-Hernández, L., Meléndez-Ortiz, H. I., Cortez-Mazatan, G. Y., Vaillant-Sánchez, S. and Peralta-Rodríguez, R. D. (2022). Gemini and Bicephalous Surfactants: A Review on Their Synthesis, Micelle Formation, and Uses, *International Journal of Molecular Sciences*, 23, 1798.

Heakal, F. E.-T. and Elkholy, A. E. (2017). Gemini surfactants as corrosion inhibitors for carbon steel, *Journal of Molecular Liquids*, 230, 395-407.

Huang, R. T. W., Rondla, R., Wang, W.-J. and Lin, I. J. B. (2017). Gemini imidazolium salts comprising Cl^-, BF_4^-, PF_6^-, $AuCl_4^-$ counterions: Synthesis, thermotropic liquid crystal study and use of $AuCl_4^-$ salt precursor to AuNPs, *Journal of Molecular Liquids*, 242, 1285–1295.

Ilincă, T. A., Pasuk, I. and Cîrcu, V. (2017). Bis-imidazolium salts with alkyl sulfates as counterions: synthesis and liquid crystalline properties, *New Journal of Chemistry*, 41(19), 11113–11124.

Kamal, M. S. (2016). A Review of Gemini Surfactants: Potential Application in Enhanced Oil Recovery, *Journal of Surfactants and Detergents*, 19, 223-236.

Kapernaum, N., Lange, A., Ebert, M., Grunwald, M. A., Haege, C., Marino, S., Zens, A., Taubert, A., Giesselmann, F. and Laschat, S. (2022). Current Topics in Ionic Liquid Crystals, *ChemPlusChem*, 87, e202100397.

Kaur, G., Kumar, H. and Singla, M. (2022). Diverse applications of ionic liquids: A comprehensive review, *Journal of Molecular Liquids*, 351, 118556.

Kowalczyk, I., Pakiet, M., Szulc, A. and Koziróg, A. (2020). Antimicrobial Activity of Gemini Surfactants with Azapolymethylene Spacer, *Molecules*, 25, 4054.

Kumar, S. and Gupta, S. K. (2010). The first examples of discotic liquid crystalline gemini surfactants, *Tetrahedron Letters*, 51, 5459-5462.

Lehmann, M., Jahr, M. and Gutmann, J. (2008). Star-shaped oligobenzoates with a naphthalenechromophore as potential semiconducting liquid crystal materials?, *Journal of Materials Chemistry*, 18, 2995-3003.

Lei, Z., Chen, B., Koo, Y.-M. and MacFarlane D. R. (2017). Introduction: Ionic Liquids. *Chemistry Reviews*, 117, 10, 6633–6635.

Li, P., Reinhardt, M. I., Dyer, S. S., Moore, K. E., Imran, O. Q. and Gin, D. L. (2021). Effects of structural modification of (alkyldiene-imidazolium bromide)-based gemini monomers on the formation of the lyotropic bicontinuous cubic phase, *Soft Matter*, 17, 9259-9263.

Li, X., Bruce, D. W. and Shreeve, J. M. (2009). Dicationic imidazolium-based ionic liquids and ionic liquid crystals with variously positioned fluoro substituents, *Journal of Materials Chemistry*, 19, 8232–8238.

Menger, F. M. and Littau, C. A. (1993). Gemini surfactants: a new class of self-assembling molecules, *Journal of the American Chemical Society*, 115, 22, 10083–10090.

Noujeim, N., Samsam, S., Eberlin, L., Sanon, S. H., Rochefort, D. and Schmitzer, A. R. (2012). Mesomorphic and ion conducting properties of dialkyl(1,4-phenylene) diimidazolium salts, *Soft Matter*, 8, 10914-10920.

Pană, A., Iliş, M., Micutz, M., Dumitraşcu, F., Pasuk, I. and Cîrcu, V. (2014). Liquid Crystals based on silver carbene complexes derived from dimeric bis (imidazolium) bromide salts, *RSC Advances*, 4, 59491–59497.

Robertson, L. A., Schenkel, M. R., Wiesenauer, B. R. and Gin, D. L. (2013). Alkyl-bis(imidazolium) salts: a new amphiphile platform that forms thermotropic and non-aqueous lyotropic bicontinuous cubic phases, *Chemical Communications*, 49(82), 9407-9409.

Trilla, M., Pleixats, R., Parella, T., Blanc, C., Dieudonne, P., Guari, Y. and Chi Man, M. W. (2008). Ionic Liquid Crystals Based on Mesitylene-Containing Bis- and Trisimidazolium Salts, *Langmuir*, 24, 259-265.

Wu, Y., Tan, H., Lin, L., Guo, H. and Yang, F. (2022). First Gemini-type fluorescent ionic liquid crystals: Synthesis, fluorescence emission and self-assembly in liquid crystals and gels, *Dyes and Pigments*, 206, 110667.

Yang, M., Stappert, K. and Mudring, A.-V. (2014). Bis-cationic ionic liquid crystals, *Journal of Materials Chemistry C,* 2(3), 458–473.

Yuan, F., Chi, S., Dong, S., Zou, X, Lv, S., Bao, L. and Wang, J. (2019). Ionic liquid crystal with fast ion-conductive tunnels for potential application in solvent-free Li-ion batteries, *Electrochimica Acta*, 294, 249-259.

Zhang, Q., Shan, C., Wang, X., Chen, L., Niu, L. and Chen, B. (2008). New ionic liquid crystals based on azobenzene moiety with two symmetric imidazolium ion group substituents, *Liquid Crystals*, 35, 1299–1305.

Index

I

K

L

M

P

S

T

U

V

X